Livability and Sustainability of Urbanism

Village of Yorkville Park, Toronto, Canada 2016

Bagoes Wiryomartono

Livability and Sustainability of Urbanism

An Interdisciplinary Study on History
and Theory of Urban Settlement

Bagoes Wiryomartono
Toronto, Canada

ISBN 978-981-13-8974-0   ISBN 978-981-13-8972-6   (eBook)
https://doi.org/10.1007/978-981-13-8972-6

This Palgrave Macmillan imprint is published by the registered company Springer Nature Singapore Pte Ltd.
The registered company address is: 152 Beach Road, #21-01/04 Gateway East, Singapore 189721, Singapore

We know each other from faces but hardly want to talk. We greet each other and forget with day and night. We belong together but attack each other with mouths and fists. Despite it all, we stay together because of our burning passion for proximity and fear of alienation: urbanism.

To all and nobody

Introduction

This book is an interdisciplinary scholarship of urbanism in the context of humanities and social sciences. The work comprises theoretical and empirical explorations of urban livability and sustainability. The main idea of this book is to divulge and unfold the concept of urbanism from various aspects of humanities and social sciences. The purpose is to broaden and deepen the understanding of the concept of urbanism in dealing with the necessity for the livability and sustainability of the urban life-world. Substantially speaking, the study is a theoretical investigation on the meanings and roots of urbanism based on historical phenomena, literary sources, personal observations, and lived experience. The intensive observation of urban phenomena for this study was compiled and consolidated with the author's lived experience of practice and teaching in the field of urban planning and development for more than three decades.

Methodologically speaking, the study's approach is phenomenological, which is characterized with the exposition, divulgence, and dismantlement of relations between phenomena and concepts in the context of the urban life-world. Accordingly, the relationship between concepts and happenings becomes crucial in understanding what are the things in the context of the world. In this worldly framework, the things of urbanism are spatiotemporally in relation to each other; the outcome and content of study on the relationship between conceptual things and their factual happenings of urbanism transcendentally are subject to lead our understanding back to the worldly context. This is the way of identifying and recognizing our experience of the things of urbanism. In other words, theory in this study is understood as the knowledge of observational experience while practical apprehension is the knowledge of lived experience. The purpose of the study is to bring into the light the urban phenomena that comprise the ideas, forms, structures, functions, and elements of communal settlement. Moreover, this study is to set up the foundation of urban theory which is based on the hermeneutic investigation. In doing so, the urban theory is not only a narrative of observations and excerpts of practical experience but also a hypothetical system of philosophically founded propositions which is constructive for an interdisciplinary platform for social sciences and humanities.

Even though the question of urbanism is philosophical, its nature is empirical from daily experience. The question is considered resourceful for the instigation of most human endeavors because most of them take place in the urban settlement, such as arts, architecture, literature, sports, music, cinema, and entertainment. The question of what urban is prior to any other question associated with urban phenomena. The significance of the question lies in its conjectural capacity to provide urban development with a theoretical foundation. In doing so, we have a big picture of any action and decision on urban matters and issues. From this provision, the theoretical investigation is necessary. In doing so, theory as circumspect knowledge from observation finds its locus on the table of discourse.

In this book, urban livability is understood as a phenomenon of happenings that gather people, things, and domains in the specific spatiotemporal context of the city; this context is the life-world of urbanism. Meanwhile, sustainability is conceived of as the capacity of urbanism that enables people to cultivate their sociocultural and economic existence and development without the depletion of their current resources in the future. Moreover, phenomenology is understood as the way of seeing things according to their presence in space and time. The presences of things are demonstrated, conceived, understood, and conserved by people with concepts, habitations, customs, and traditions. The phenomenological study is intended to unfold and divulge the things that construct the interrelations of their meaningful and functional whole within the existential system of the life-world. Phenomenology of urbanism is an attempt to excavate and dismantle the aspects of the urban reality from various relevant traces and resources—from geographical location, history, trade-interaction to philosophical insight.

The focus of the whole studies is a cross-disciplinary endeavor to unveil, unfold, and dismantle the relationships between concepts and phenomena of urbanism in terms of the life-world. The exposition of this relationship enables readers to understand the reality of urban habitation as the unique life-world of a society. In this book, urbanism is conceived of as a system of thoughts, efforts, forces, and intentions that work, develop, maintain, and sustain for a sociopolitically organized settlement. Accordingly, urbanism is inseparable from the sense and essence of development so that people are able to establish and grow their sense of home.

The whole book contains two parts with theoretical chapters and pertinent case studies. Under the notion of home, urbanism is understood as a system of dynamic relationship between people and places that happen among things in time. The theoretical explorations of this book include the disclosures of fundamental concepts that matter for urban settlement and its phenomenal existence in the context of polity, society, historicity, design, and development. Meanwhile, the empirical expositions encompass the case studies from the Malay Peninsular world and Toronto, Canada.

Part I consists of seven chapters. All these chapters concern theoretical exploration of urbanism. Chapter 1 is a philosophical reflection and meditation of urban place. Even though building and living are usually understood as routine happenings, the question concerning what and why the place is has been forgotten. Places are transformed and understood historically, but they are the settings and contexts of human populations. To what extent are such places human? Chapter 2 is about formative concepts of urbanism, citizenship, and civil society.

In this study, urbanism is understood as a conceptual system of habitation comprising ideas, efforts, forces, and intentions within the framework of a sociopolitically organized settlement. Accordingly, urbanism is in the search for the sense and essence of development in the context of the built environment. Chapter 3 excavates the lesson learned from the historical Greek Polis after Aristotle's Politics. In history, cities and towns have been always equal with the centers of civilization. The focus of this chapter is to examine ancient Greek concepts, which are essential and constructive to the idea and reality of urbanity. What is planning? The question leads the Chapter 4 to unfold and uncover the plans that matter for urbanism. The focus of this chapter is community planning by which the site of the urban life-world is laid out on the land for various land uses and activities. With a systemic construction of land uses and activities, Chapter 5 is to explore the question of how such diverse functions of land are able to create safe, healthy, and attractive places. This chapter explores the function and contribution of urban design to urbanity with three-dimensional forms and domains. However, all we have done with planning and design approach for the built environment needs a set of guiding principles toward livability and sustainability. Chapter 6 is concerned about the constitutive elements and function of environmentally friendly urbanism that matter for livability and sustainability. This chapter explores the aspects of environmental sustainability for land use planning and urban design in terms of public policy. History is a considerable resource for intervention in the making of the built environment. Chapter 7 discusses the issues and transformations of the human condition in the Global Age that matter for urbanism. This chapter explores the theoretical elucidation of global urbanism from the reality of the globally connected worlds. The theoretical exploration in this chapter is constructed in confronting the sense of home. To what extent does the globalization challenge the necessity for feeling at home in an urban context?

In Part II of the book, chapters are organized to unveil case studies concerning the reality of urbanism. Chapter 8 is a study concerning the everyday life of urbanism in the West Malay World. This chapter attempts to find the characteristics of urbanism from daily experience to historical and sociocultural roots that have shaped, established, and developed an urban tradition in the Malay world, and how such a tradition has been living and being sustained until today? In Chapter 9, the historical exploration is continued and focused with Malacca as a case study. This chapter is considered as an in-depth study of waterfront urbanism in Southeast Asia, which is historically notable for the concept of *bandar*. In Chapter 10, the exploration is focused on the relationship between political culture and urbanism in West Malaysia in the context of planning and design intervention. How does political power work that plans, builds, develops, and sustains urban settlement in the Malay Peninsular world? In dealing with globalization, this study is necessary to elaborate its scope of exploration concerning global awareness worldwide. This exploration leads to the phenomenon of intentional awareness of urbanism in the global context. Chapter 11 is an attempt to divulge the relationship between mind and urbanism that leads to global awareness and reality, with the City of Toronto in Canada as a case study.

Pedagogically speaking, the aim of this book is to examine, scrutinize, enrich, and enlarge theoretical understandings on urbanism, as well as to explicate distinctive aspects of urbanism from a specific locality to global connectivity. In doing so, urban studies are able to bring philosophical principles and everyday habitations of urbanism that construct, develop, and sustain the sense of home for people in terms of urban society and community. There is nothing the same in urbanism but the necessity for a place people call it home. It is the place where people find their sense of stay because they feel safe, comfortable, and excitement in a socio-environmental context. They have each other but not necessarily know each other by the first name. They belong together but not necessarily feel being bounded by a social contract. They share something in common without having lost their sense of privacy and individual uniqueness. And, they experience the complexity and diversity of places, things, and people but not necessarily ignore and disregard simplicity and integrity. Urban society constitutes the sense of home with their sociocultural activities while urban community brings about the feeling at home with their socioeconomic livability toward sustainable growth. All of these—livability and sustainability—are necessarily put in terms of equal opportunity and social justice for all.

In doing so, the essence of urbanism is the human endeavor and cultivation of holistic understanding of their presence in the world.

Notwithstanding, the problems, issues, and matters of urbanism have been explored and studied by many scholars and specialists elsewhere, but it is necessary to build a comprehensive study on the integration of history and theory into a systematic body of knowledge. In doing so, the relationship between historical evidences and theoretical findings is able to be unfolded and uncovered within an interdisciplinary context and understanding. As matter of fact, the question of urbanism is indispensable and insistent for an urban educator, planner, designer, policymaker, and developer in their works because it is about the sense and essence of development in manifold ways. More precisely, the question is about why do people build, grow, and maintain to work, live, and play in the city? This study attempts to bring back the question of urbanism into the context of livability and sustainability in a sociopolitically organized settlement.

Last but not least, 'Urbanism, Livability, and Sustainability' is expected to be among the most comprehensive and profound books on urbanism of our time. Substantially, it is considered rigorous and sharp in its investigation, logic and imaginative in its investigation, and nevertheless accurate, deeply self-reflexive, and imaginative in its presentation, as well as appealing for far-reaching understandings on phenomena of urbanism. Moreover, this book is distinguished from the available books in publication by its interdisciplinary and philosophical approach to the complexity of urban phenomena. The exposition and explication of these phenomena is considered indispensable that enable and engage readers to understand the reality of urban habitation as the life-world and home.

Contents

Part II   Empirical Exploration

About the Author

Dr. **Bagoes Wiryomartono** is independent scholar and currently on leave Faculty Member at the School of Design, University of Pelita Harapan, Karawaci Tangerang, Indonesia. Bagoes earned his Doctorate in architecture and urbanism from Aachen University of Technology in 1990 in Germany. He was a postdoctoral fellow for architecture at the East West Centre, Honolulu, Hawaii, and Fulbright scholar at the Smithsonian Institution, Washington, DC, USA. The area of specialization of his teaching and research experience is focused on history, theory, and design of urbanism of various cultures and traditions in Southeast Asia and North America. He was a senior lecturer at the Bandung Institute of Technology (1981–1983, 1993–2002) and visiting research associate at the Asian Institute, University of Toronto Canada (2003–2005). He was associate professor at the Department of Architecture, Faculty of Built Environment of Universiti Teknologi Malaysia (2010–2013). Most recent scholarly books by Bagoes Wiryomartono include: *Javanese Culture and the Meanings of Locality: Studies on the Arts, Urbanism, Polity, and Society* (forthcoming 2016). Lanham: Rowman & Littlefield. *Perspectives on Traditional Settlements* *and Communities: Home, Form and Culture in Indonesia* (2014). New York-Berlin-Singapore: Springer Business Media. He published his research in various Scopus journals on history and theory of architecture, urbanism, and culture in Southeast Asia as well as philosophical studies within phenomenological movement.

List of Figures

Part I

Theoretical Exploration

Fig. 1 Sudirman Golden Triangle, Jakarta 2014 (Photograph by author)

1

Ontology of Urban Place

Urban Place and Location

Under the notion of ontology, this study understands the things from the relationship between their concepts and historical presence in certain geographical context. Regarding concepts are contextually and historically related to each other; as a concept, place never stands alone as a socially isolated entity, for example, house: The concept is never understandable without owner, inhabitant, location, event, construction, form, material, etc. This interrelationship constructs systems or sets of the ecosystemic wholeness; the wholeness of a house is a place to stay for a family of certain people in a certain geographical location and certain period of time. Concepts related to place—such as house, park, lake, market, shop, office, and school—are within the web of concepts that constructs the life-world. In other words, a place is a phenomenon social and geographical context that constructs the platform of reality for human habitation. Ontology of urban place is the theoretical construction of understanding what and why is the appealing location for urbanism important; this is in the context of the life-world as a transcendental and meaningful system of wholeness.

© The Author(s) 2020
B. Wiryomartono, *Livability and Sustainability of Urbanism*,
https://doi.org/10.1007/978-981-13-8972-6_1

Culturally speaking, domains and districts in the urban area are conceptually functional which are articulated with the attributive, denotative, connotative, and adjective property, such as residential, industrial, recreational, commercial, and multifunctional. It is not by accident that the category of urban places and domains reflects and represents the idea and value of purpose and preference as well as characteristic and identity. Domains are differentiated and distinguished according to their use, value, location, and dimension. However, beyond and behind the concepts of the domain, there is an ideological system that works to determine their presence sometimes in a contradictory way, for example, urban and rural, town and country, global and local, etc. Some domains are conceptually articulated without oppositional category but simply a denotation of use, characteristic or cipher such as park, square, street, road, path, and trail.

Urban Place in Space-Time

Space and time are indisputable archaic phenomena of *being* that construct the historical context of a thing; both are physically indefinable, infinite, and a priori in its reality. Space and time are a priori an archaic phenomenon because without both being and beings are impossible for experience. The question concerning space-time has been occupied philosophy from Western tradition since ancient Greek thinker of atomism Since Democritus (460–370 B.C.). He understands the concept of space-time as the infinite, in which a particle moves or something is. Plato perceives space-time in Timaeus (48e4) in the context of *chora*; for him, *chora* is neither being nor nonbeing but the interval, in which the look of something gives or happens. Aristotle (384–322 B.C.) in Book IV Physics elaborates the question of space-time in terms of *topos* as the site of happenings; this is considered a reinterpretation of Plato's *chora* in Timaeus within the experienceable context. Accordingly, Aristotle departs his discourse concerning space-time from the question of what it is; for him, space-time is the categorical components of *topos*, from which something must be somewhere spatiotemporally; *topos* is the site of happenings for something because there is *energeia*. The processes of transformation in the universe or cosmic nature are made possible because of this *energeia*. According to Aristotle, *energeia*—power, effort, energy, capacity, and potentiality (Beere 2009, 33)—works within the cyclic system of *aitia*, in which what is received or gained as trust or gift should be given back with responsibility or reward as thankfulness (Heidegger, *Basic Writings* 1977, 289–91). Aristotle's doctrine of *aitia* brings about cyclic process of *energeia* in full circle, in which the relationship between trust and responsibility as well as between presence and goodness (*t' agathon kai to ariston*, Nichomachean Ethics 1904a18–23) comes into terms; this goodness is experienced as *eudaimonia*. Aristotle calls the full circle of *energeia* as *telos*; the notion of *telos* is often understood as finality but actually, it is a fulfillment of cyclic process of *energeia*.

The elaboration on space-time as category has been the work of Immanuel Kant (1791/1988) in his *Critique of Pure Reason*, Part I *Transcendental Aesthetics; Space-Time*, and *Transcendental Idealism* (Kant 1781/1990). Accordingly, space-time is a priori structure for the categories of the reality, presence, and possibility of things. Kant does not explore the relationship between place and space-time, but Heidegger does in the context of clearing. Heidegger describes space-time as the constitutive structure for the world in the sense of closeness or proximity, which is experienced in way of dealing with everyday comportment (Heidegger 1927/2010, Chapters 22 and 24). Space-time for Heidegger is not ever an extension of the thinking subject, but it is constitutive for the human existence; in Heidegger's thought, existence means standing out into the truth of being. Thus, space-time is already there prior to the existence of man as the site of the truth's happening. Nevertheless, the spatiotemporality is not thought by Heidegger as a matter but a clearing, which is discovered and experienced when the phenomena of closeness and orientation-direction come into being. Heidegger's clearing recalls us to Plato's *chora* that it is already there prior to any happening. As matter of fact, the possibility of *being* is categorically conditioned by the space-time awareness. Hence, the human existence is spatiotemporally conditional because the *being* of man is impossible without space-time (Ray 1991, 135).

Furthermore, a place is not a random area or arbitrary spot because it is ontologically associated with a destination of happenings about people and things. A thing is not a random matter or event, but it is something intentional that attracts and draws parts together as a system or a wholeness. Thus, a place is the site of things that implies it is the spatiotemporal location, condition, and position where and when the things happen. Questioning what and why a place is is indispensable to the quest of space-time in relation to the certain destiny of happenings. The question leads us to investigate the relationship between humankind and their destiny in space-time framework. Architecturally speaking, the space-time is the endless source of any possibility of providing people and things with places. However, Aristotle reminds us in Metaphysics on *aitia* that any gift contains responsibility. The generosity of space-time that provides people and things with the potential realm of happening—Plato calls this potentiality as *chora*—comprises the responsibility of giving back with goodness and beauty. Aristotle understands this responsibility as the consequence of the cyclic transformation of *energeia* in terms of *telos*. Since everything through the transformation of energy takes place in space-time setting and context, place in terms of Aristotelian *topos* is the site of happenings with the quality of full circle or *telos*. In this sense, place is not any domain or location but the site that brings about the goodness according to its ontological destiny. Under the notion of ontological destiny, urban place is the site of urbanity or being *urbane*, in which people live, work, and play together to do their best and goodness as their thankfulness to what the resources have been given for their presence.

Concerning its space-time destiny, the urban place is not merely the location of settlement for certain number of populations and productions. Rather, the origin and the sense of urban place pertain to the purpose of community that is experienced as collective sense of purpose. The destiny of urban place cannot ignore the human call for the collective pursuit of growth in various fields. In order to understand the destiny of urban place, one needs to recognize the ontological boundary of town or city; this boundary is not a physical matter, but it is the collective mental states and affairs that spiritually drive and bring people together toward decisions and actions according to the principles of being *urbane*.

Architecturally speaking, the principle of being *urbane* for the built urban environment is the availability of safe, healthy, convenient, and attractive public domains for social interactions and gatherings. Accordingly, the urban place is a plastic artwork, in which the sense of its whole presence is experienced when its public realms are vibrantly occupied by the crowds. Heidegger signifies the sense of the plastic art as the embodiment of the truth of being in the work in its bestowing places (Heidegger 2007, 13). The truth of *being* in this sense is experienced by people as the integration of their presence and the place in infinite series of the now. Heidegger speaks of the happening of this truth in terms of *Ereignis* that the place gathers historically the fourfold of the worldly phenomenon; the fourfold encompasses the whole mortal and immortal properties, as well as the earthy materiality and heavenly spirituality. For Heidegger (2012), this happening unfolds and divulges the difference between withdrawing-preserving and emerging-revealing; in this event, humankind and being befit each other in a deep sway of being in sameness.

The place is not any spot (Locke 1841/2007, 97–100; Creswell 2014, 66–7). The place is characterized with a particular set of the occupied area with name or designation. Any appropriate designation of place is set up and fabricated to accommodate certain activities and collective interests. Urban places, in the context of public realms, are subjects to sociopolitical ideology (Mitchell 2003, 128) because they are the domains of the communal life-world of diverse populations, in which contestations, collaborations, and negotiations of power come into play. In democratic countries, such dynamics reveal and conceal the intentionality of urban collectivity so that people enjoy their freedom but sink into anonymous crowds.

As a matter of fact, the space-time of a location has its own destiny in terms of use and function in the whole system of urbanism from metropolitan, township, region, district/ward, subdivision, precinct, neighborhood to compound. The sense of place lies in its manifestly unfolded locality (Massey 1994, 151–3). Accordingly, locality comprises demography, topography, layout position, geographic orientation, property, architectural feature, and landscape characteristics. The locality of urban place is demographically characterized by highly concentrated populations, geographically situated in a complex layout of settlement, and a network of utilities. The locality of a place is established with a collective memory of the populations, which supports its life-world.

However, the sense of local place is not merely the site of happening for occasional encounters. Rather, the locality of a place is the outcome of historical cultivation of events. The rituals and regular festivals enhance and enrich such cultivation with collectively shared memory among the populations. The phenomenal events constitute the structure of personal and collective memory (Jordan 2003, 31–8) that matter for the establishment of local history and the sense of home. Being away from such a place will recall the sense of longing for its people. The relationship between people and places unfolds and uncovers the destiny of home in the cosmological realm. This destiny is recognized through familiarity and habitation. This experience builds up the identification of people with their place. A self-identification to place is nothing personal but the sense of belonging to a community and settlement. Such a self-identification brings about the space-time intimacy of people with their place in terms of home or stead. The memory of places recalls people for their gratitude to what has been given by space-time to their existence and growth. Establishing memorial structures and naming of places after certain persons are the way to express the gratitude of community to people and place, which are historically associated with the space-time locality of the area. Space-time provides people with the connection of people to their built environment. Space-time is inseparable from what it *is* now. Something happens is always to take space and time into experience. Being as presence correlates one's space-time to 'the right now' that brings about personally lived experience; the experience is based on the series of moments that comprise the unity of when and where into one occasion.

Urban Place and Aesthetics

Regarding space-time quality, public domains are the architectural things with the spatial permeability, accessibility, and fluidity for human movements. The permeability of public domain is made possible because of its spaciousness and connectivity with paths and other domains. Meanwhile, the accessibility of a public domain includes safe and convenient accesses by foot, strollers, and wheelchairs. However, as architectural works, public domains need to be in the network of public paths and areas so that they are not isolated from public accessibility for function and fire safety. The space-time quality of public domains provides people with the opportunity of leisure and relaxation with visual interactions and communications.

Fig. 1.1 Chartres Street at Jackson Square, New Orleans, USA 2014 (Photograph by author)

Aesthetically speaking, attractive public realms are provided with a well-designed form of landscape and fixtures that befits specific designations, such as for park, square, court, garden, trail, or pedestrian easement.

Environmentally speaking, public realms are the outdoor domains with healthy atmosphere and amenities. In this sense, public domains are necessary to integrate in a pedestrian network that ensures the green areas that potentially hold and naturally filter pollutants out of rainwater prior to flow into rivers or aquifers. Healthy atmosphere of public domains is necessarily provided with ample sunlight, less noise, and fresh air circulation. Well-designed amenities provide public domains with conducive built environment for social interactions that help to establish a strong community. Such amenities include fixtures and public facilities for safe and convenient social gatherings and interactions. The aesthetic experience in public places is the pleasure of seeing and being seen as well as the joy of casual and impersonal nearness with others. Despite proximity, the delicate sense of intimacy in public domains remains impersonal and nameless. Installing sculptures, fixtures, monuments, water elements, and stairs helps the public places to enrich and enhance the aesthetic experience of visitors and passer-by (Fig. 1.1).

Urban Place and Territoriality

Defining territory for public domains is neither to claim property nor to isolate area. Rather, defining the boundary of open areas is to prevent the domains from the unexpected uses such as for crimes and dumping wastes. The necessity for the territorial demarcation of public domains is not only for public safety and health but also for the management of human actions in respect of the sustainability of nature (Sack 1986, 58–61); this includes use of green elements and trees for defining spatial boundary. Indeed, the visual control for public safety is necessary either actively by people on spot or official camera. However, by design, public places should be visually controllable so that potential unwanted accidences are architecturally prevented. In order to achieve this, the spatial affiliation of the public places should be designed clearly so that people in the surrounding buildings are able to casually observe what happens in the adjacent areas. The notion of territoriality, in this case, is more than just the physical boundary of the public places, but it is about social inclusivity of the areas into the community's control. In other words, public places are not no-man's lands, but they belong to the adjacent community.

The establishment of territoriality for public domains is socially possible through the habitation of regular users. Habitation and domicile are not practically the same. Both have a different space-time approach; this concept of habitation relies on familiarity while the notion of domicile is about claiming or controlling land. The concept of domicile stems from the Greek word *domos* meaning domain for home (Massey and Finch 1998, 463); this concept has been adopted and elaborated with the Latin word '*dominus*' meaning lord or master. The territoriality of public areas by habitation is not in terms of domination so that its open realm is necessary impartial form any vested interest, but it belongs to the collective civic interest and responsibility. Regular visitors and users should be part of civic guardians for the openness of the public domains. In order to maintain the openness, public places are necessary regular maintenance and replenishment. In doing so, the territorial claim by regular users or visitors is practically prevented and avoided.

The ancient Greeks understood *topos* as the site of happenings and to experience beings as a whole (Slomkowski 1997, 43–71). Accordingly, *topos* is a percentage of space with the specific designation, which is associated with the gathering of beings. The sense of *topos* lies in the disclosure of space-time for certain activities. Therefore, *topos* has a boundary made known with its destiny that allows mortals to stay for experiencing the right now of engagement. The sense of intimacy comes into play in this experience; the experience liberates people from their routine to empathetic delight. The aesthetic experience of people to enjoy their public realm is the joy of engagement for sinking into anonymous crowd. All of these remind people to Charles Baudelaire's *flaneur* and Benjamin's *Arcade Project* observing the urban life-world of Paris. There is a pleasure for mingling and dissolving oneself into the anonymous crowd in public realms of the city. This aesthetic experience unfolds the joy of impersonal proximity with others.

Despite its anonymous crowd, the public realms of the city are not expected to be a jungle but the open domains with territorial boundary. Place marking is not found randomly across landscape but rather is an ordered component of socially constructed space (David and Wilson 2002, 8). The territorial boundary of public realm is necessarily recognizable so that people feel safe of their whereabouts. Regarding the sense of territoriality, a public place is not a leftover area, but it is necessary to be designed with particular form with specific designation, characteristic, capacity, modality, and relation to other places. Formally speaking, public domains, such as squares, parks, parkettes, courts, and gardens, are necessary to be institutionalized with official names.

Urban Place and Organization

Place is the site and location of happenings, which involves people. Accordingly, a place never stands for anything but human events. House, neighborhood, village, town, city, and nation are nothing but places with manifold characteristic and capacity. Originally speaking, the place is the site of a whole of beings where people are in their closeness to each other and to the things. A place is characterized by the proximity of people and other beings that is incorporated in various forms, structures, styles, and organizations. In daily activities, the manifold capacity of the place shows its designation and properties such as town, village, and house.

The village is commonly associated with a settlement with estate encompassing several family houses in the rural area. Architecturally speaking, a village is not densely populated; the built-up area for buildings occupies the land less than its open environment with a significant proportion. However, a village is distinguished from a hamlet or group of houses because of its administrative status. In other words, a village is a politically organized community less than the basic requirement of populations for a township or county. Nevertheless, a village is traditionally known as a community with strong social cohesiveness. In such a community, people know each other by their surname and understand each other based on customary codes and mores. Unwritten norms and practices belong to collective rights and responsibility which are set up and developed within a village community.

The physical borderline between villages is commonly established with natural boundaries or normative agreements. Economically speaking, rural village community is commonly associated with agricultural production. As a rural community, the village is economically subsistent, even a self-contained system for daily sustenance. However, there is a village community in the urban area that has nothing to do with agricultural production or farmstead. The notion of village in the urban settlement has a different connotation; the notion of 'village' in this context is usually associated with a historical background. To a certain extent, the concept of the village in an urban area is applied for highlighting and articulating the idea of the pedestrian scale of the streetscape with lively gathering and moving crowd. The essence of village settlement lies in its living community with strong traditional values and practices. Such an urban village sets up and develops certain rituals, customs, and traditions that bring about their own collective pride and identity among other communities.

In sustaining the community life, the rural village is provided with amenities where the congregations and gatherings take place. Though the permanent marketplace is not typical for a village, the weekly market is a part of village economy life. In an urban area, village community is formally associated with a subdivision. The ancient notion of such a subdivision is quarter, which is originally from the Roman colonial garrison town.

The Roman city is spatially divided by historic *Arcus-Decumanus* axial streets into four areas: northeast, northwest, southeast, and southwest. Historically speaking, one-quarter of the town was occupied by populations with specific profession, class, rank, and status. Subdivision of town or city usually does not develop a social affinity because of its inhabitants' mobility and individualism of urban lifestyle.

The town is institutionally man-made places with a concentration of heterogeneous population, which make their living from manifold activities of proximity, engendering, and service. In contrast to the village, the town has always a permanent marketplace where proximity for exchanges takes place. Among other human places, the town is a man-made environment that manifests as the phenomena of proximity for dwelling in manifold forms and associations. Settlement as the form of proximity for a dwelling has its inherently reproducing capacity that enables its existence developing. A town in its essential form is an institutionally defined boundary of settlement that is able to sustain and to develop its resources for exchange with other towns and regions. Characteristically, the town is provided with the permanent marketplace, spiritual hall, town hall, and public open space or square. Town has a center of social, economic, political and cultural activities so that urbanity is set up, developed, and cultivated in its fine and matured form and language.

Economically speaking, urban areas are supported by diverse populations with specific skills and knowledge of productions and services. The diversity of professions and resources brings about its contribution to the urban life with socioeconomic and political activities that sustain a dynamically cooperating system of production and service for the common welfare. In dealing with the complexity of its resources, possibilities, and potentialities of interactions, town and city are established as a political institution with written rules and regulations. Such rules and regulations are in need to establish order, right, responsibility, and mutual understanding among its citizens. All this is about safety, security, and well-being of the community. Regarding its resources, diversity of its inhabitants, and complexity of interactions, the town grows and expands its boundary and intensity.

The physical growth of town is indicated by its economic and population growth. Large town is a city. It is a place where dynamic diversity and complexity of socioeconomic, cultural, and political activities are managed within the institutionalized system of government. In contrast to town, the city is more bluntly secularized place for the various way and lifestyles. City population is mostly divers in its origin, beliefs, education, occupation, and lifestyle orientation. In a city, there is hard to have a single system of values. Diversity and multivalent lifestyle belong to urban living condition because such various possibilities are necessary for the development of civilization. Diversity provides the city with the multifarious possibility of interaction that leads to the process of learning for each other toward the development of knowledge and arts. Although conflicts and struggles become to be the part of urban life, innovations and inventions have its fertile ground in the city life.

Concluding Notes

There is no place without local embodiment. The sense of locality lies in its capacity to create, establish, develop, and sustain homegrown identity and character. Locality is a historic cultivation of populations, who live, work, and play there. Building places is not simply erecting and establishing physical structure, but it is building the life-world that builds and sustains the uniqueness of the relationship between people and their location. Concerning sustainability, the way people deal with their places is never indifferent but ethically responsible and accountable in a way of being grateful to their locality. This is not simply a moral imperative but an existential obligation because the sustainable locality provides people with the sense of home. The ritual gatherings for the local festivals are the ways how the locality is unfolded, unveiled, and identified as the site and boundary of their life-world. In doing so, people find their abode in the places they build. All this leads people to find and dwell at their home on the earth under the sky. Urban place is a man-made environment for human settlement with highly concentration of activities that provides people with the domains of stay, play, and work for civility.

In the Age of Global Information, the ontological question concerning urbanism remains primordial and the same as it was in ancient times that is to deal with the sense of home for the political community. This is the sense that what we build, develop, inhabit, and maintain is able to fulfill our basic, social, political, and spiritual need as a person and a member of society. The sense of home is tied to the feeling of safe, secure, being convenient, and being to give us the ability to deal with any given situation. The question leads us to deal with the contemporary world that challenges our sense of home in any aspect of the experience. The global life-world does not only take place in the built environment, but also on the cyber platform and network. To what extent is the ubiquitous concept of urbanism still valid and relevant? The proliferation of cyber connectivity underpins the disappearance of the public sphere and the reinforcement of private intimacy. Even though cyber social media, such as Facebook and Twitter, are publicly accessible, such platforms are arguably unfit to establish and replace the institutions of the public sphere meant for political discourse. These technological improvements are not to diminish and dismantle the processes and productions against the presence of public realms, such as the virtually constructed self-casting and the addiction of being attached to virtual connectivity through personal gadgets and computers.

References

Beere, Jonathan. 2009. *Doing and Being: An Interpretation of Aristotle's Metaphysics Theta.* Oxford: Oxford University Press.

Creswell, Tim. 2014. *Place: A Short Introduction.* Oxford: Blackwell.

David, Bruno, and Meredith Wilson. 2002. *Inscribed Landscape: Marking and Making Place.* Honolulu, HI: University of Hawaii Press.

Heidegger, Martin. 1927/2010. *Being and Time.* Edited by Dennis J. Schmidt. Translated by Joan Stambaugh. Albany NY: SUNY Press.

———. 1977. *Basic Writings.* Edited by David Farrell Krell. Translated by Adolf Hofstadter. San Francisco: Harper & Row.

———. 2007. *Die Kunst und der Raum.* Frankfurt am Main: Vittorio Klostermann.

———. 2012. *Bremen and Freiburg Lectures: Insight into That Which Is and Basic Principles of Thinking.* Translated by Andrew J. Mitchell. Bloomington and Indianapolis, IN: Indiana University Press.

Jordan, Jennifer. 2003. "Collective Memory and Locality in Global Cities." In *Global Cities, Cinema, Architecture, and Urbanism in a Digital Age*, edited by Linda Krause and Patricia Petro, 31–48. Fredericksburg: Rutgers University Press.

Kant, Immanuel. 1781/1990. *Critique of Pure Reason*. Translated by Mortimer J. Adler. London: Encyclopedia Britannica.

Locke, John. 1841/2007. *Essay Concerning Human Understanding*. Oxford: Oxford University Press.

Massey, Doreen. 1994. *Space, Place, and Gender*. Minneapolis: University of Minnesota Press.

Massey, Gerald, and Charles S. Finch. 1998. *The Natural Genesis*. Baltimore: lack Classic Pers.

Mitchell, Don. 2003. *The Right to the City, Social Justice and the Fight for Public Space*. New York: Guilford.

Ray, Christopher. 1991. *Space, Time, and Philosophy*. New York: Routledge.

Sack, Robert David. 1986. *Territoriality and Its History; Human Territoriality*. Cambridge: Cambridge University Press.

Slomkowski, Paul. 1997. *Aristotle's Topic: What Is a Topos?* Leiden: Brill.

2

Urbanism, Residency, and Society

The Quest for Urbanism

Historically, the theory of urbanism has been indebted to the works of Patrick Geddes (1915), Louis Wirth (1938), Lewis Mumford (1961), Kevin Lynch (1981), Spiro Kostof (1991), Nan Ellin (2013), Short (2014), and Scott and Storper (2015). Their works are not only historically significant for the education of generations of urban planners and designers but also fundamental for our theoretical understanding of urban development. Nonetheless, since the beginning of the 1960s, the practice of urbanism in North America has been profoundly criticized for its fiasco and decline in providing public realms with safe and vibrant public space (Jacobs 1961; Rae 2005). The dichotomy of the urban and suburban is crucially widening that challenges new approaches for reconciliation. The emergence of New-Urbanism movement has compelled and renewed traditional approaches for urban practice in which the sense of community has its weight to be addressed and incorporated by planning and design (Katz et al. 1994; Passell 2013; Fulton 1996; Talen 2013). The central issues for contemporary urbanism have to deal with the liveability

© The Author(s) 2020
B. Wiryomartono, *Livability and Sustainability of Urbanism*,
https://doi.org/10.1007/978-981-13-8972-6_2

of the public realm and environmental sustainability and the economy as a whole.

Recently, the theoretical debates and discussions concerning urbanism have been well documented and compiled in response to various aspects; from the urban processes, urban forms to urban societies (Kelbough and McCullough 2008; Waldheim 2012; Ellin 2013; Branigan 2002; Ramirez-Lovering 2008; Binne et al. 2006; Calthorpe and Fulton 2001). The contribution of this compilation evidently sheds the light on the need for a comprehensive and systematic body of knowledge in urban theory. The more specific and broader aspects of urbanism are explored and unfolded, the more worthwhile and urgent is the necessity for a fundamental and philosophical theory on urbanism. Epistemologically, urban theory is necessary to deal with three areas of humanity: knowledge, ethics, and aesthetics.

As a systematic knowledge, the urbanism is necessary to respond to the question of what urbanity is that conserves and sustains the quality of the urban life-world; this quality is sociologically associated with communally regarded and respected behaviors. In terms of urbanism, urbanity is everything that establishes, upholds, and protects peaceful, healthy, and productive interactions and exchanges. As knowledge, urbanism is necessary to set up a systematic body of know-how on urbanity that leads people to strive for the best in terms of productions and services. In this point, urbanism provides public policies, regulations, and guidelines toward sustainable improvement and growth for economy, culture, and environment. All this is necessary to be in the framework of highly concentrated activities of habitation. Accordingly, the challenges of urbanism are to set up, develop, and update the criteria of the public excellence based on the constraints and potentialities of locality.

The ethical framework of urbanity is founded on the mutual respect of people and the cultivation of excellent endeavors. The written codes, regulations, and policies are apparent forms and ways that enable urbanism to work on practical way. However, there are some unwritten comportments, mores, customs, traditions, and manners, which are practically exercised for urbanity. Such informal codes of conduct are not lawfully organized, but they are communally imperative in terms of collective decorum. As a matter of fact, urbanity is historically cultivated from the necessity for respect, fairness, harmony, and exquisiteness. Respect is considered inherent in human soul that prevents and keeps out humankind from rudeness, improper conducts, and impudent deeds.

Environmentally speaking, urbanism concerns the question of proper relations among beings, in terms of human interactions between themselves, as well as between them and their environment. Sustainability is not merely economic but also environmentally imperative for urbanism. Accordingly, rethinking and redefining technology is crucial that brings a synergy between economic growth and environmental conservation. Nonetheless, urbanism is inevitably to keep the entropy of development at low level so that it ensures the production and services with less wastes and pollutions. Regarding the conservation of environment, urbanism is ecologically in concert with the natural processes of energy transformation with less hazard leftovers. The sense of urbanism for environmental sustainability lies in its design and technology so that all materials used for urban development are recyclable and reusable without burdening and encumbering public health and comfort.

Aesthetically speaking, urbanity is about the good quality of city living condition; this is demonstrated with the state of affairs of safe, secure, healthy, convenient, comfortable, and beautiful experience. All these indicators are provided with steady sustenance that pertains to the notion of urban. The word urban stems from the Latin words *urbanus*, *urbs*, and *Urbis*, meaning city, city form, and citizen, respectively, that articulate the prevalently civilized world (Zijderveld 1998, 11–14). The sense of urbanism entails the capacity for gathering resources for the development and improvement of human endeavors through the works of arts, science, and technology. The endeavors are nothing but the whole cultural achievements that bring forth the transformations of raw resources into the openness of beings. This is the events when urban people experience their sense of home within the proximate system of collective dwelling.

The sense of collective settlement is experienced by people in their gatherings in open places. In such gatherings, urbanity takes place such as on the side streets, side boulevards, halls, squares, marketplaces, gardens, and parks. All these places enable people to interact with each other visually and directly. The necessity for the existence of public realms is inevitable and essential for urbanism because it is the happenings when and where the urban life-world comes into being (Simmel 1950; Arendt 1958/2013; Habermas 1989; Sennet 1977); without this existence, people do not experience the sense of living, working, and playing in urban areas.

Socioeconomically speaking, urbanism cannot ignore the existence of the permanent marketplace and political authority. Despite its complexity, exchanges and political order become the essential components for the existence and sustenance of the urban life-world. As a matter of fact, urbanity is not only a highly abstracted concept but practically actual in terms of socioeconomic interactions in the public realms of the city. Such interactions entail any form, style, manner, and way of proper communications, fair negotiations, and secure transactions. In this point, governmentality is the essential components of economic exchange and social order. Accordingly, the market is not only socioeconomic activities but also—in the broadest sense of the words—a legitimate trading institution. In this sense, the market is a public institution for various exchanges, deals, and negotiations with accountability and fairness. The phenomenon of permanent market is historically the origin site of public interactions and socialization that establish, develop, cultivate, and sustain the socioeconomic exchanges of civil society. The exchanges include not only the trades of goods but also works, skills, knowledge, and ideas. The municipality is the political institution of urbanism that guards, keeps, maintains, and protects people together as a civil society. In return, taxation is considered imperative as the form of civic responsibility of people to participate in the collective agreement and the guardian of public order and security.

What is the town or city? We are not going anywhere with the question before being able to define and formulate what urbanity is (Andranovich and Riposa 1993, 1–3; Steinberg and Shield 2008; Zijderveld 1998, 21). To appreciate urbanism is inevitable to include urbanity. Without this, urbanism is worthless. What is then being *urbane?* Being *urbane* is the concept of articulating decent manner or well-mannered in contrast to rude and impolite behavior. The root of urbanity is urban which is derived from the Latin word *urbs* meaning city life (Gates 2003, 2). Urbanity is originally the quality and the state of being—living, working, and playing—of people in the urban settlement; this is demonstrated with its healthy, convenient, conducive, and safe environment for manifold human activities of being and making. Urbanity concerns civility which is shaped, developed, established, and sustained by urban citizens in terms of ethics, laws, manners, and lifestyles that cultivate, uphold, refine, and uplift their human dignity. Urbanity is more than simply a set of moral virtues. Rather, the sense of urbanity lies in the experience of being free from any danger and that of the joy of home in the city.

The call for urbanity is the sociopolitical, economic, and cultural necessity of urban populations. Urbanity is a social phenomenon that happens mostly in various gatherings of people in public places. In the broadest context, urbanity is demonstrated by people in their social interactions in urban areas based on mutual respect, care, hospitality, and friendliness. In order to achieve this quality of life, urbanism needs to set up, establish, develop, and sustain a conducive built environment and a system of ethics, manners, and practices through regulations, laws, and educational guidelines. Educational facilities and public services are important institutions for establishing, transferring, and training the virtues of urbanity.

Urbanity is the quality of city living condition that is indicated with safety, security, health, convenience, comfort, and beauty. All these indicators are provided that peace has been maintained at its steady sustenance. Urban is town characteristics originated from Latin words *urbanus*, *urbs*, and *Urbis*, meaning city, city form, and citizen, respectively, that articulate the prevalently civilized world (Zijderveld 1998, 11–14). The essence of urbanity lies in its capacity to gather resources for the development and improvement of culture through the works of arts, science, and technology. Civilization is nothing but the whole cultural achievement made by urban citizens that brings forth the raw resources into the openness of beings throughout human works. It is about the transformation of nature to culture.

Urbanity takes place mostly in the public realm in various places: street, boulevard, hall, square, marketplace, and park. All these places have their own specific role and function in shaping, developing, and sustaining urban activities. The notion of the public realm is not simply a legal matter, but also metaphysically necessary. Historically speaking, urbanism cannot ignore the existence of the permanent marketplace and political authority. Despite its complexity, exchanges and political order become essential components of urbanism. As a matter of fact, urbanism is not only a highly abstracted concept of civility concerning socioeconomic interactions in the built environment, but it includes any form, style, manner, and way of communication and interaction, such as in cyberspace. However, both institutions are the essential components of urban society. The market is not an only socioeconomic institution but also—in the broadest sense of the words—a trading institution where various exchanges, deals, and negotiations take place. It is the origin site of public interaction and socialization that establishes, develops, cultivates, and sustains civil society. This includes trade in goods, products, works, skills, knowledge, and ideas. On the other hand, authority is the institution of service that guards, keeps, maintains, and protects people together as a civil society.

Needless to say, in the course of history civility as well as urbanity has been misinterpreted in various ways. And, the urbanity evolves more and more complex and diverse in activities, products, and services. However, civility of community remains the most important thing of what and why people build the place. It is neither about numbers of population, nor intensity of buildings. It is about the quality of relationship among beings based on respect and synergic mutuality for sustainably growing together dynamically and beautifully. Civility reflects the basic need of human beings in terms of the community that they are not simply biological beings, but cultural and spiritual ones.

Today, the sense of urbanity has been displaced and commercialized to private realm in terms of shopping malls, theme parks, private halls, casinos, fitness centers, and clubhouses. The nature or urbanity as public gatherings in public places has been challenged and compelled by excessive consumerism to follow the logic of commodity of the free market economy. In this situation, urbanity without economic values could be perceived as the non-productive social activism of welfare state with taxpayers' money.

Essentially, somewhat urban is because of its urbanity. It is neither a moral issue nor economic matter. Rather, urbanism is political as well as ideological that works on the necessity for intensity, diversity, interconnectivity, and productivity of human settlement. The need for intensity is demonstrated by the phenomena of a habitation with proximity, interaction, and exchange with a permanent spot. This intensity is enhanced and enriched by the phenomena of the diversity of populations and associations. This diversity potentially creates new demands that challenge innovations and improvements. However, intensity and diversity without the phenomena of interconnectivity are idle if all resources and activities are not well connected together as a productive and sustainable system. In addition, the phenomenon of interconnectivity is to bring about intense activities and diverse potentialities purposeful and useful as a whole.

Characteristically, populations in the urban area are diverse in the division of labor and sociopolitically organized as productive and servicing units. In short, we call this organization as civil society and not community (Cohoone 2002, 12). Why do people organize themselves as a society? The urban settlement is characterized by the proximity of settlement and diversity of division of labor, as well as state polity and exchange of commodities. Since the sense of urbanism is cultivation for civility, all these characteristics are potentially constituent and formative for urbanity. We can say that civility is the intentionality of urbanism. This phenomenon is experienced as something that unconsciously drives and directs human actions toward respectful and considerate behaviors.

Civility is epistemologically derived from Latin word *civitas*, meaning a form of political and social participation for cultural cultivation. Lesson is learned from Aristotle's Politics that society is not simply a group of people with the certain goal to achieve, in the name and for the sake of civilization. Rather, society is the concept of a politically organizing system for human socialization, communications, and interactions, which are stewarded by laws, customs, and institutions. In doing so, they are themselves respectable and decorous citizens. Such a system establishes, protects, and serves all urban citizens with certain rights and responsibilities for public safety, order, and peace. Nevertheless, Aristotle points out that the bond of this urban society is justice because only social fairness provides urban society with equal opportunity and professional ethics for all. Based on this, best ideas and works are competitively made possible.

Today social justice in most urban areas in North America is still the work in progress. Disparities in income, education, and occupation are everywhere. It is obviously not because of the lack of law and order, but mostly because of misunderstandings, misperceptions, and prejudices against minorities, based on race, gender, sexual orientation, and financial situation. Marginalization of citizens for many reasons could lead to a suicidal process of urbanism because it creates disparities and dysfunctions of societies as a whole. As a political entity, urban citizens need to work together for their urbanity. The existence of functional societies is vital for urbanism because these professional, business, and interest groups or associations create, develop, elaborate, and support specific interests and achievements for urbanity. Leaving certain people on the sideline of urban associations is letting them down and against urbanity.

Despite its various political systems of government, human comportment is the central issue for a living and growing together in a permanent location because dwelling is actually building a relationship with the place, people, and time that we call familiarity, vicinity, and affinity. Without such building, we never come to terms with what inhabitation is. The urban settlement is a Greek *topos* in Aristotelian Physics Book IV. It is a frame of reference for arrival and departure. And it is not an empty space but a locality with a contextually interwoven network of communications. Urban site is a *topos* of state polity and organization with a permanent market as its core of socioeconomic interaction and socialization.

From ancient Greek *polis*, we learn that the daily life in the urban settlement was mainly focused on the public gathering at *Agora*. The Greeks built the *agora* for exchanges and social interactions. It was the place where urban citizens—*polite*—gathered and cultivated their political, social, cultural, and economic life (Hansen, Symposium August 1997, 61). Lack of strict theological doctrine in ancient and classical Greek gave the room for philosophy to occupy the necessity for metaphysics that brought about *agora* not simply place for economic and social exchanges but also for interactions of thought. Socrates was mostly around at *Agora* for asking philosophical questions to every citizen he met. He deliberately entertained people with his dialogues with other guest thinkers from other cities.

The phenomena of urban settlement refer to the exigency for the cultivation of human behaviors and endeavors supported by land cultivation. The manifold professional specializations, guilds, social associations, crafts, and artworks are the manifestations for such urban cultivation. Only in relation to human sociocultural development, land cultivation and its resources have the sense of purpose for civilization.

Urban Settlement

The concentration of settlement becomes an apparent form of urban form. However, without a locally established seat of political authority, a concentration of settlement remains in the vicinity with a large number of populations that do not have its own household in dealing with its surrounding resources. The compactness of settlement is likely to have something to do with the necessity for efficient collaboration; proximity enables people to have intensive interactions and direct communications. The physical nearness is considered as a human condition as social being so that people are able to overcome their fear of loneliness. The nearness is the state of affairs when people feel that everything is within the reach of their hand so that they do not need to worry about something. Being with others in physical nearness enables people to grow together biologically, socially, emotionally, and spiritually. House, town, and city are nothing but the outcome of the necessity of people for being with each other in nearness.

Architecturally and geographically speaking, urban settlements mostly occupy strategic locations for safety, security, and access to resources. This phenomenon is apparent with respect to accessibility and manageability of land for trade so that its commodities are within reach for exchange with other regions. Unsurprisingly, the relationship between trade and state is essential for urbanism. The state does not only provide trade with political shelter but also with regulative and statutory protection. In return, trade is due to pay revenue to the state on a periodic basis. The origin of urbanism is impossible detached from the history of state and trade because town and city are the sites of where the consolidation of authority for exchanges and communications takes place. Heidegger (1927/2010, 94–6) describes the phenomenon of the nearness of a dwelling that binds together toward freedom. As a matter of fact, human settlement is an existential place where people have a chance to be free from loneliness and alienation. It is an existential place because people are able to grow and cultivate their potentials and resources as a community.

Moreover, the political stability of the state is the steward of trade and urbanism. It does not matter what kind of political system of the state is—monarchy, oligarchy, or democracy; healthy and fair trade for various commodities is vital and essential for urbanism. Then, we can say that the existence of political control and permanent market are essential institutions of urbanism. Historically and global-wide, it is evident that the emergence of urbanism is inseparable from the establishment of state polity and permanent marketplace.

Furthermore, the existence of marketplace marks out the necessity for a center that attracts and brings about the gathering of people from various directions. Center for urban activities is actually temporal happening that takes place in the various site. Happening in regular time and place constitutes a spatiotemporal identity of what center is. Based on this regular happening, a center is entitled to be established as an urban institution, such as market, stadium, school, town hall, theater, church, and shopping center. Market activity usually takes place at the most strategic location. This is with respect to accessibility from the whole area of settlement. The other centers are for religious, transportation/communication, sport, and cultural attraction. The more sophisticated and complex urban activities are, the more are the number of its urban centers. Beyond marketplace, the locations of such center are not absolutely in the midst of urban settlement. The establishment of centers in urban system constitutes the dynamic vitality of urbanite existence because this place potentially draws most people to give their best. Interactions and competitions, as well as conflict resolutions, take place in such centers. Urbanity exists meaning that it stands out for the excellence.

However, the essence of urban entity lies in its possibility for the cultivation of mind and body of societies that brings forth the civility of urban citizens which are characteristically based on local resources, customs, and traditions. Locality and the sense of the place are not architecturally given but culturally developed and elaborated which are in regard to the existing condition and historical precedence of the vicinity. Accordingly, every site and settlement group are unique because of their local nature and culture. Even though urbanity is universal in respect of its values on civility, customs, and traditions, as well as the physical and historical environment, play an important role to build and establish its urban flavors, characteristics, features, and spheres.

City and town are nothing but urban phenomena and concepts that represent everything that matters about urbanity. Both concepts, city and town, are essentially identical but with the different category in terms of area, intensity, complexity, and diversity of populations. However, town and city have their common ground as an urban settlement. In the Canadian context, the city is an urban settlement larger than the town. And, an urban settlement is inhabited by more than 2000 populations. Of course, the density of populations plays a decisive role in characterizing a settlement group as urban or non-urban habitation. However, acceptable density varies in North America between 386 and 1000 populations/square kilometer (Roberts et al. 2005, 89). Since the concentration of populations generates interactions, communications, and exchanges, the large-scale settlement is necessarily provided with various facilities and services that enable people living in the area to sustain their daily nourishment. Given that living for a human being is not simply sustaining life with food and drink, human beings are complex in nature because they have the will, and mutually dependent with each other because they as well as other living beings are by nature members of the community.

In ancient time, the Greek *polis* had integrated urban and territorial control into a single management system of state (Pomeroy et al. 1998, 84–9). Then, we learn that *polis* is politically an independent city-state. Exploration and exploitation of natural resources, as well as agriculture, become a long-standing tradition of territorial empire since ancient time. Urban areas become the spots of economic exchanges. Controlling territorial regions where farmlands and urban areas are the way of ancient empire since Sargon of Akkad,-later Babylonia ca 2400 B.C.-, accumulating their wealth and power. Sumerian cities like Isin, Larsa, Ubaid, Eridu, and Ur under Sargon were not simply cultural centers as achieved by Sumerian civilization (Kipfer 2000, 349), but also administrative centers of its agricultural regions.

Historically, the emergence of urban culture came about from the phenomena of the concentration of populations at the river valley such as in Ancient Sumerian civilization from circa 3000 to 3200 B.C. The traces of urban area are recognizable from the remnants of foundation for highly concentrated settlement and drainage canalization. The relationship between urban settlement and agriculture is ostensibly shown with the knowledge and technique of irrigation. Water management experience plays important role in the structure of urban settlement. However, the contribution of agriculture to urban settlement is obvious for sustaining the surplus of the domestic economy of crops.

In most ancient and historical urban sites, archaeologists found the remnants of three historic things: coinage, pottery, and inscription. Urban settlement and culture have something to do with these three valuable things which are associated with the currency trade and seal of authority, household equipment, literacy, codes, and laws. As a politically organized system, the urban settlement is not only about building structure and infrastructure, but also a group of societies that live and interact with coinage, pottery, and scripture. Coinage is not only the means and form of currency for exchange, but it is also a symbol of authority that signifies trade with the formal, standard, and legal means of transactions.

Besides its functionally economic sense, coinage is a sign of civility in trade activities. The importance of coinage for urban culture is essentially based on the fact that trade and communication have an inherent relationship which is deeply rooted in the need for civility. In other words, coinage is not simply a currency or medium of economic exchange, but it is the sign of interpersonal value that seals the deals of commodities exchange and reimbursements of service. Economically speaking, currency provides people with the feeling of certainty and security in their transactions.

Besides coinage and pottery, script in pictographic form is widely found in ancient urban sites of Mesopotamian, Egyptian, Indus, China, and American civilizations (McIntosh and Twist 2001, 7). What is the relationship between urban settlement and script? The script is actually the system of signs and symbols that represent information and message of bit or aspect of the world. The emergence of scripture in various forms, matters, and ways of articulation gives us something to know or to learn. The necessity of script for urbanity is not only for public communication, but also to seal the deals, in respect of partnership and treaty. Messages or information, which is subject to be known publicly, is necessarily brought into interpersonally understandable form if its representations are inscribed or carved into a system of signs and symbols. Drawing and picture play a decisive role in conveying messages or information with the presence of their author. Thus, script represents something communicative related to the thought and feeling of its author.

The emergence of the script in urban settlement might have mediated something about norms, principles, codes of law, and agreements that are publicly accepted and kept as conventionally fixed references of conducts and behaviors so that people interact, communicate, and trade on the peaceful and trustful ground. The meaning of script is also indispensable for trade (Bogucki 1999, 333–4). Moreover, the emergence of script enables the authority to make a public announcement concerning regulations, codes of law, and policies.

The urban settlement is established and developed on the ground of human development in understanding what exchange is in the broadest sense of the word. The establishment of exchange comes about with a socioeconomic system based on laws and orders. In order to make exchange possible, there is the necessity for currency that represents and carries out the mechanism of interaction, transaction, and cultivation. In this point, the currency of human relationships—in a social context—is respect to each other, whereas, in the economy, the currency of exchange is money as the representation of market value. The relationship between urban settlement and the existence of permanent market lies in the fact that economic exchange requests the conventional agreement of value for any good and labor. The market has its own wisdom to determine the value of good and labor based on the relationship between supply and demand that establishes and sustains the productions, consumption, and services in the urban system.

Respect for others is unaccountable and unreliable if it does not have any conventional reference. Since human beings are communicative, there is a necessity for convention based on language. Nevertheless, oral communication for intensive interaction and exchange is not accountable and reliable for making an agreement. There is a necessity for filing what has been agreed with a system of signs that is to inscribe everything understandable, conventional, and referential for dealing with agreement and conformity. Laws and orders need such an inscription so that its messages and principles become reliable and accountable for keeping interaction, transaction, and instruction clear and trustworthy.

Writing and literacy are nothing but the basic media of communication that bind people together within the language of proximity and familiarity. The importance of writing and literacy for urban life lies in its memorable guardianship message through time. Without writing, the cultivation of human endeavor and achievement is hardly attained with efficient effort. The advent of writing is actually in response to the necessity for the cultivation of knowledge and experience as well as for confinement of contracts inefficient way. The necessity for cultivation and confinement comes about in urban life in dealing with the fact that human being is not simply biological being. Rather, the being of man is always in the search for the open region of being that is to bring forth everything resourceful as well as every relation to others to be the things within the framework of cultivation.

Writing and literacy come into active players in the development of urbanity since the ancient civilization. Regardless of the language used for its system of signs, there is barely urban settlement without writing a record for sales, loans, affidavits, labor contracts, formations, and dissolution of partnership (Bowman et al. 2000, 976). Historically, writing is preserving and conserving what is spoken as well as thought and felt within the system of signs. Urban dwelling comes to terms with writing in order to preserve and conserve what has been cultivated so that learning process comes about a generation to generation with the efficient matter, form, and way of deliverance and sustenance.

Nonetheless, the contribution of writing in keeping agreement trustworthy is one important point for urbanity. However, the origin of writing is the condition of a human being that is communicative and associative. The necessity for preserving and conserving communication and association through writing is actually the essence of sustainable cultivation on a permanent site. Urban fabric is to set up and to establish the permanence of place, whereas writing is in a similar way to record exchanges, commitments, pacts, and happenings of the world. Building and text come to terms of a settlement that establishes and sustains the phenomenon of home.

Nevertheless, the advent of writing is not adequate for the establishment of the urban settlement. There is still the necessary authority to keep the integrity of place and the world within the system of the human community. Building and script do not work as the existential structure of the home in regard to its being in time. The exigency of the authoritative institution comes about in response to the call for the guardian of the world of exchanges and interactions. Since the world is the understandably interrelated system of beings based on language, relations among beings are necessarily organized and maintained within the framework of the sociopolitically ordering system.

The exigency of authority for ordering system comes into being in urban society as a whole. It is in response to the call for the sense of organizational purpose. Since any community has a purpose in dealing with the earthly given resources, urban dwellers never have the sense of purpose for reproduction and production without regulation, occupation, and direction. Then, the establishment of urban authority in terms of municipality comes about to bring forth urban society as a whole reproductive and productive in dealing with their world as a home system. In order to maintain and sustain the authority of regulation, occupation, and direction, there is an exigency for a tribute to those who sit and hold authority. We understand such tribute as a tax. Thus, the relationship of taxation and urban settlement is indispensable from the mutual reciprocity between the cultivator and conductor.

Nonetheless, the urban economy does not work without taxation, because it is the essential mechanism of urban household that maintains and sustains regulation, occupation, and direction of production and reproduction in the broadest sense of the words. The existence of taxation is essential to establish the elaboration of cultivation from land to human possibility under a sociopolitically organized system that conducts regulation, occupation, and direction. Manifold forms and ways of taxation do not change its essential sense as tribute from cultivators to their conductor as well as protector. Taxation comes about at the first time in agricultural cultivation such as in ancient cultures in Egypt known as *scribes*, in Greece land as *eisphora*, in Roman Empire as *portoria* as well as in any sites of urbanity on the globe (Thomsen 1964, 11).

The relationship between urban settlement and taxation is in the interest of well-established organization of production that sets community and its protector within the system of mutuality. Athenian democracy was probably the oldest exemplary system of taxation that the relationship between cultivators and conductors was mutually established with the direct and common interest of urban citizens. Beyond Athenian democracy, taxation was the subject of the tribute of land and cultural cultivators to their emperor as well as a landlord that conducted and protected the system of cultivation. Besides political stability, the rise and fall of civilizations are indispensably dependent on the management of cultivation of land and its trade through taxation. Monuments, citadels, and palaces are never out of commission from taxation.

Currency, script, and taxation are three components of the urban system. They work together as urban household or economy. The existence of the three urban components enables transformations in urban settlement within the framework of the socioeconomically organized system so that the sense of growth and development comes into being in various matters, forms, and manners of human endeavor and comportment. Nevertheless, currency, script, and taxation are not only economic, but they are essential for the cultural sustenance of urban society as a whole. They bind all people as urban citizens in a mutually interrelated system. Currency, script, and taxation work together as functionally circulating and opening out the mechanism of exchange for social, economic, cultural, and political life.

Without flow and growth in the context of exchange, urban life does not have any sustainable future because its productive transformation does not bring all inhabitants into a mutually collaborating system based on the diversity of interests. The occurrence of such collaboration in urban life takes place in the urban center, especially in the market. All this does not work well without a permanent site for exchange (Catanese and Snyder 1979, 33; Sanders and Price 1968, 98; McClain and Wakita 1999, 38). The establishment of the permanent marketplace for urban settlement is not simply the place for the exchange of commodities and transactions. Rather, the establishment of the market is the institution of meeting in a public realm that engenders urbanity on a daily basis. Accordingly, it is the sign of openness and fair trade between demand and supply that drives and moves economic productivity. The importance of the market for urbanity lies in its open and public characteristics that build and sustain trust and reliance of traders under the protection of state law and order.

In respect of its architecture, the urban settlement is unique in terms of its relatively compact and concentrated structure. The concept of town is originally from Old English *tun* and Old High German *zun* meaning an enclosure or a defined boundary or a physically enclosed domain for a living (Banhart and Steinmetz 1988, 582). Regarding this phenomenon as an enclosure and architecturally defined boundary, the sense of town lies in the necessity for proximity within a physically secured boundary. Is this enclosure a territorially conquered land of man over nature? Historically, an open or fortified settlement is not decisive to constitute the concept of what an urban entity is. Forms, matters, and ways of what urban are manifold, but the construction of domain is essential. Even though the walled town or city is not a common case, the tendency to settle with proximally built form is evident.

Several centuries before Anno Domini, the architecture of urbanism of citadels and fortified structures has been traceable as a constructed domain in ancient sites of Sumerian city of Ur or Tell al-Muqayyar, Acropolis of the Greek Athens, and Rome of the Roman Empire. The city wall had been functioning as the guardian of ancient civilizations. The wall was not only a defense structure but also the symbolic borderline between the urban life-world and the non-urban one. Within the architecturally walled domain was the place where legal codes and manners were exercised and enacted. Such architecturally fortified structure is still available in European towns. Some citadels incorporate castles, such as Brest Fortress of Belarus, Ark of Bukhara in Uzbekistan, Tower of David in Jerusalem, Citadella of Hungary, Cittadella of Gozo in Malta, Hamina Fortress of Finland, and Kremlin of Moscow in Russia. Today, several citadels were demolished, but the ruins remain, such as Acra of Jerusalem, Cadmea of Greece, Erbil of Iraq, and Liege of Belgium. Some citadels remain, however, solitaire without township (Blackwood 1822, 346). Commonly found, the cities or towns with citadels were mostly constructed at the river plains with the castles at the height overviewing the whole landscape. Even though some towns and cities in Southeast Asia do not show to have such a sturdy town wall (Tarling 1999, 131), the relationship between township and the domain within the architecturally fortified domain is most evident. In South East Asia, the building materials for the fortification were mostly not durable enough against the tropical climate. Nevertheless, the remains of limestone bricks and stones are found in the ancient sites of Langkasuka in Old Kedah, Malaysia, Sriwijaya in Jambi Sumatra, Indonesia, and Majapahit in Mojokerto Java, Indonesia.

The ancient Greek Agora and Roman Forum Trajan were not simply the places of gathering but the esplanade and honorific platform for citizens with provision and precaution; both places were decorated with a whole series of the fine portrait statues of eminent personalities (Dillon 2006, 102; Bravi 2015, 135). Well-mannered comportments were practiced and honored by people in the public places of the Greek and Roman civilization. In ancient and Hellenistic Athens of the fifth century B.C., one could not distinguish ordinary residents from slaves from their appearance, but from the way they behaved and spoke. Although outspokenness was highly regarded among Athenian citizens under Pericles' democracy, well-mannered temperament and comportment belonged to the decorum of citizenship (Kagan 1991, 59).

Governmentality and Public Sphere

Governmentality is about the art and management of powers, interests, and influences. More importantly, governmentality concerns a system of management and control over the access to resources, in terms of labor, skills, material, fund, and land. The governmentality underpins the strategy and practice of introducing, organizing, maintaining, performing, challenging, reviewing, consolidating, and supervising policies concerning public matters and affairs. Contemporary studies on governmentality are indebted to the works of Michel Foucault in the context of power relations; Foucault argues that within the framework of agonistic platform of the political life-world, the ancient Greeks invented a unique power relation which is coined by him as subjectivation—the relation of oneself to oneself (Foucault 1985); this subjectivation is an ethical awareness of the self for being able to be free from any kind of domination by others. The concept of subjectivation pertains to the capacity of the self as master of itself and author of self-recognizing problems in the context of considerate freedom, *eleutheria* (Aristotle, *Politics, Volume IV*, 1997; Aristotle and McKeon, *Introduction to Aristotle*, 1992, 1263b; 1317b, 10–11). Subjectivation entails the self-awareness of freedom so that one is able to manage his/her conduct in public realm according to law and order based on the ontological necessity for civil liberty. The concept of the city in

Ancient Rome as the settlement of civitas and city was equivalent with the Greek politês and polis in the Ancient and Classical Greece around 600–300 B.C.

The ancient Greek experience of *polis* teaches us that urbanism is a sociopolitically organized system based on the use of discursive language, *logos*, through debates, deliberations, negotiations, and persuasions instead of that of force with violence and coercion. All this is the human condition that humankind is sociopolitical being; in Aristotle's word, humankind is *zoon politikon* (Aristotle, *Politics, Volume IV*, 1997, 1253a, 7–38). The sense of political things lies in their relation to public interests and public domains. Being engaged and part of public issues, constraints, and opportunities are in human condition. Hannah Arendt (1958/2013) characterizes this human condition as the striving for being human that comprises works, labors, and actions; this threefold is coined by Arendt as *vita activa*.

As activities, *vita activa* are potentialities that make humankind to be human; the essential thing of this human condition lies in the fact that humankind is *bios politikos* (Aristotle, *Politics, Volume IV*, 1997). Arendt's human condition is about a human *being* that enables them to work and grow together as an urban community for their cultivation. All this is made possible because they have language. Language is not just for interactions and communications, but it is human disposition and capacity that enable them to set up and establish the conceptual formulations and discourses concerning social ordering system in written form, *nomos*. Because of human linguistic competence, humankind is able to identify, organize, manage, and operate beings within the socially organized system of happenings we call it the life-world. The spatial and circumstantial realization of this human condition in the context of urbanism is the emergence of the public sphere as the urban life-world; this is the essential realm of human necessity as a political being.

The ancient Greeks understood the public sphere in terms of the *polis* as a realm of freedom and presence. Habermas observes that in European history, this Hellenic public sphere has been handed down and transformed from its original meaning—as the realm of civil liberty—to the status attribute of power and legitimacy in the context of bourgeois society (Habermas 1989, 1–26). This transformation has diminished the significance of freedom and transparency in the urban space. The public realm of the *polis* is ideologically inclusive for all citizen. Even though the concept of citizenship in ancient Greek *polis* does not include women and slaves, the intentionality of the public realm is to underpin the reality of civil society as a politically free and present group of populations. In order to establish this civil society, democratic governance is structurally in need, so that competitive environment for Aristotle's *koinoniken arêten* (Aristotle, *Politics, Volume IV,* 1997, III.13 1283a, 38–40) is potentially established, developed, and sustained.

The concept of urbanism in terms of Aristotle's political community—*polis*—is not impartial from the necessity for social interactions that take place in public realm based on wise conducts and practical wisdom in terms of *phronesis* (Mcclelland 2015, 64–6). This concept is not about the wisdom of knowledge, but it belongs to human decisions and actions based on respect and being considerate of others. The close notion with the Greek concept of *phronesis* is civility, well-mannered behaviors and conducts in the public realm. Important sources on ancient Greek governmentality come from the studies of Jean-Pierre Vernant (1983, 212–34), Mogens Herman Hansen (1991, 1997), and Aristotle's Politics by Kraut and Skultety (2005). Accordingly, *polis* is a sociopolitical community with a self-governing system with its center of public realm at Agora. *Polis* reflected the most important thing for the Greeks, namely individual freedom. Even though slavery was part of the existence of *polis,* among the Greeks, they shared the sturdy beliefs in the worth of individual freedom. Old Spartan jokes referred to the fact that in daily Athena they were not able to see the difference between citizens and slaves.

During the Classical period (500–338 B.C.), more than 1200 *polis* existed throughout the land of Hellas. Accordingly, as an independent sociopolitical entity, the *polis* was not simply a city-state form of urban settlement. Rather, the *polis* was properly the life-world of the independent self-governing community for religion, government, and trade with agricultural support from its surrounding rural areas, villages with farmlands. Membership in a *polis* was represented by a male person who had the right to participate in making policy through public meeting at the Agora. Indeed, women and children, as well as slaves did not have the political right. Nevertheless, respects to the laws, to the family, to the gods, and to the *polis* are due to all Greeks for what good is. The thing that united the Greeks was their myths and respect to individual freedom of man within the concept of public good.

Under the notion of civility, urbanism is the intentionality of the socially cultured world that directs, attracts, drives, and strives people for the public good through various works, labors, and activities. This is nothing but human cultivation in the context of political community. Civility is the outcome of human endeavor as a political community that is associated with the framework for their cultivated works, socioculturally refined behaviors, and activities that are practiced, codified, and developed in a human settlement with highly concentrated populations. Civility is considered as the essential grain of urban development. Civility has been historically developed with written codes and laws for social orders, currency, and tax for trades and businesses, and governance for authority and administration.

The civil world is the reality of urbanism that is sociopolitically constructed with civic systems consisting of institutions, societies, and places. Global urbanism transforms these systems from local embedment to open and loosely mobility. In the urban context, civic activities take place in the spaces between buildings. Since ancient times, the marketplace has been playing a significant role in shaping and developing urban societies and communities. Various interests, hobbies, and skills build and establish associations and professional groups that we call urban societies while socioeconomic cooperation and network set up communities. Societies and communities work hand in hand with the establishment and sustenance of urban settlement.

Society and Citizenship

Urban citizen in Latin is called as *civic* or *civitas* from which the word city originates. Under the notion of *civic* or *civitas*, we understand citizen or member of the community with political right to the city public sphere (Miles et al. 2004, 205–7). The manifold concepts related to *civic* include civil (proper and polite citizen), civility (politeness), civic (municipal, city), and civilization (well-mannered behavior within a socially sophisticated system). *Civitas* was a notion of sociopolitical privilege for urban inhabitants, but it is not applicable for women and slaves. Today, this concept has been redefined as a citizen, meaning a person with the right to the city and state affairs. Speaking of *Civitas* is describing urban citizens that live as sociopolitical beings based on laws and orders. With the privilege as *Civitas*, urban inhabitants had the right to participate in making a political decision through senators as their representatives.

To be a good citizen, a Greek man was due to have the fourfold of goodness mentioned above with proportional credence in him. They just developed codes and laws based on common sense for daily life, and coinage for their exchanges. In contrast to former ancient Babylonian civilization, ancient and classical Greek *polis* was under despotic rulers and not a multicultural political community like in Sumerian and Babylonian civilization. The Greek city-states were supported by a political community of *koinônia* under the governmentality of Athenian democracy or Spartan oligarchy.

For the ancient and classical Greek, the *polis* was the political life-world they understood in their own language within the framework of political community. The existence of *polis* as the state was possible through its agricultural surplus of its surrounding villages. Beyond the *polis* was unknown territory that was not their concern and would belong to other *polis* or simple the unknown or the barbarians. The concept of city-state is essentially on the ground of politically divided state organs based on interests of citizens which were mostly in three groups: aristocrats, warriors, and common citizens. These three interest groups were actually not confined to strict organizations because among warriors there were farmers and traders. However, the city-state as a well-organized community did their assembly at *agora* where their market took place. Political and economic activities took place at the same site. Assembly and political meeting played important role in ancient Greek with various modes of representation.

Under the aristocratic government, Athenian citizens were represented by a council of elders—*arcopagus*—while the nobles delegated their voice to their magistrates—*archai* (Hansen 1991, 225–6). The interplay between magistrates, chiefs of warrior, and *arcopagus* came into play in the assembly—*ecclesia*—of Athenian democracy under Pericles. These main interest groups constituted the body of city-state government in the form of assembly; predominant political groups under the interest of kinship called *phratry* were not disappeared from the political scene of Athenian democracy; they sustained their political influence among members of the assembly—*ecclesia* (Hansen 1991, 125–7).

Regardless of geographical, cultural, and historical origin, permanent marketplace belongs to the long-standing site of socioeconomic activities that establishes, develops, and sustains urbanism. Archeologically speaking, the marketplace, *agora*, was among the earliest public buildings in ancient Greek history; the first Athenian *agora* was constructed in circa 600 B.C. that marked out the establishment of the *polis* as a sociopolitically and economically self-sufficient settlement. The original meaning of *agora* is gathering place (Hansen 1997, 61) that becomes the place for public discourse, marketplace, and leisure; the presence of *agora* signifies the existence and sustenance of public sphere for the city-state, *polis*. In various traditions, market day happens on a periodic basis, at a public urban place that gathers people for trades and social interactions. The happenings of the market take place at the permanent location that constructs the collective memory of people concerning the idea of the center or core of the daily public sphere. Moreover, the establishment of permanent market underpins the sustainability of the public sphere with its natural aura and social intimacy; this experience is crucial for the cultivation of civility.

The existence of permanent marketplace, *agora*, is the determinative factor for the establishment of the urbanity of polis because of its potentiality as the generator and hearth of the public sphere for political governmentality, social interactions, and economic exchanges. Nevertheless, marketplaces, *agora*, in ancient Greek times, were the generator of economic growth for wealth and political stability of *polis*; this growth played insignificant role in the expansion of the market because states remained major economic agencies, markets were not consolidated and shallow, with high transaction expediencies, and investment opportunities were limited (Harris et al. 2015, 7–8).

As a matter of fact, the urban economy does not work without taxation, because it is the essential mechanism of urban household that maintains and sustains regulation, occupation, and direction of production and reproduction in the broadest sense of the words. The existence of taxation is essential to establish the elaboration of cultivation from land to human possibility under a sociopolitically organized system that conducts regulation, occupation, and direction. Manifold forms and ways of taxation do not change its essential sense as tribute from cultivators to their conductor as well as protector. Taxation comes about at the first time in agricultural cultivation such as in ancient civilizations in Egypt known as *scribes*, in Greece as *eisphora*, in Roman Empire as *portoria* as well as in any sites of urbanity on the globe (Thomsen 1964, 11).

In contrast to village habitation, the urban settlement is not a customary polity but a state polity that establishes and sustains highly sophisticated society and civility. Urban habitation is distinguished from the village community that is not merely based on the scale and scope of work. The urban settlement is managed as a political community based on its diverse capacities to manage and control their resources for their social, economic, cultural, and political cultivation. As a political community, the urban settlement is challenged to grow and be productive that expands and develops their knowledge, skills, and domains for new opportunities. In contrast to an urban area, a village community is to develop its capacity to maintain and sustain a homeostasis within their production system (Trojan 1984, 23), so that cultural growth and development are restricted by the number and diversity of its populations. The cultural development of the village community is mostly in a way of refinement and elaboration, instead of exploration and accumulation.

In the Age of Global Information, the gap between rural and urban area has been diminished by the Internet connectivity. The global connectivity has partly demolished the borders between geographical places and the legal construct of territory. Anthony Gidden identifies this global connectivity as the consequence of modernity; this connectivity intensifies worldwide social relations that connect geographically distant localities in such a way so that local happenings are shaped by the events occurring of kilometers away and vice versa (Gidden 1990, 64). For urbanism, global connectivity challenges the definition and inscription of territoriality for the public sphere. The consequence of this global connectivity is the crisis of the public sphere that brings about the dynamics of power relationships between the hegemony of the global capitalist economy and the territorial governmentality of the nation-state. Ubiquitously, public sphere in the urban settlement is not a geographically and legally constructed territory of nation-state sovereignty, but a conceptually constructed realm of open interactions and communication for all urban citizens; this conceptual territory is challenged by the global connectivity of cyber access and network for its inclusivity and transparency. To what extent the nation-state has the right to control and manage the cyber interaction and communication that is in alignment with civil liberty and human rights?

The Age of Global Information puts the concept of the public sphere in the crisis of its conceptually defining territory and its legally inscribing boundary. The problem of this redefining and reinscribing lies in the disposition of the public sphere as a human condition for urbanity. There are two essential components of the public sphere in the global urbanism, namely capital and information. Theoretically, both components work for the global economy under the framework of civil liberty. However, democratic governmentality is not, under any circumstance and reason, able to manage the global capital network and worldwide access to information.

Urban resident in Latin is called as *civic* or *civitas*, from which the word city originates. Under the notion of *civis* or *civitas*, we understand resident or member of the community with the political right in the city life (Miles et al. 2004, 205–7). There are manifold concepts related to *civic* such as civil (proper and polite citizen), civility (politeness), civic (municipal and city), and civilization (well-mannered behavior within a socially sophisticated system). *Civitas* is a concept of sociopolitical privilege for urban inhabitants, but it is not relevant for slaves. Speaking of *civitas* is describing urban residents that live as sociopolitical beings based on laws and orders. With the privilege as a *civitas*, urban inhabitants have the right to participation and engagement in the making of the political decision through senators as their representatives.

The concept of the city in Ancient Roman Empire was understood as the settlement of *civitas*, and city was equivalent with the Greek *politês* and *polis* in the Ancient and Classical Greece around 600–300 B.C. Accordingly, *polis* as a sociopolitical entity was a settlement system of community with a self-governing independence with its center at Agora. *Polis* reflected the most important thing for the Greeks, namely individual freedom. Even though slavery was part of the existence of *polis*, among the Greeks, they shared the sturdy beliefs in the worth of individual freedom. Old Spartan jokes referred to the fact that in daily Athena they were not able to see the difference between citizens and slaves.

During the Classical period (500–338 B.C.), more than 1200 *polis* existed throughout the land of Hellas. Accordingly, the *polis* was not simply city-state mode of urban settlement as an independent sociopolitical entity. Rather, *polis* is properly the center of independent self-governing community for religion, government, and trade with agricultural support from its surrounding villages with farmlands. Membership in a *polis* was represented by a man who had the right to participate in making policy through public meeting at the Agora. Women and children, as well as slaves, did not have the political right. Nevertheless, respects to the laws, to the family, to the gods, and to the *polis* are due to all Greeks for what good is. The thing that united the Greeks was their myths and respect to individual freedom of man within the concept of goodness.

To be a good citizen, a Greek man was due to have the fourfold of goodness mentioned above with proportional credence in him. They just developed codes and laws based on common sense for daily life and coinage for their exchanges. In contrast to former ancient Babylonian civilization with despotic rulers, ancient and classical Greek polis was not a multicultural community like in Sumerian and Babylonian civilization, but Greek *koinônia* with Athenian democracy or Spartan oligarchy.

For the ancient and classical Greek, the *polis* was the world they understood in their own language within the framework of community household. The existence of *polis* as the state was possible through its agricultural surplus of its surrounding villages. Beyond the *polis* was unknown territory that was not their concern and would belong to other *polis* or simple the unknown or the barbarians. The concept of city-state is essentially on the ground of politically divided state organs based on interests of citizens which were mostly in three groups: aristocrats, warriors, and common citizens. These three interest groups were actually not confined to strict organizations because among warriors there were farmers and traders. However, the city-state as a well-organized community did their assembly at *agora* where their market took place. Political and economic activities took place at the same site. Assembly and political meeting played important role in ancient Greek with various modes of representation.

Under the aristocratic regime, Athenian citizens of ancient Greek *polis* were represented by council elders called *arcopagus*, whereas aristocrats delegated their voice to their magistrates. The interplay between magistrates, chiefs of warrior, and Arcopagus came into play during the Athenian democracy under Pericles. These main interest groups constituted the body of ancient Greek city-state government in the form of the assembly called *ecclesia*. Nevertheless, predominant political groups under the interest of kinship called *phratry* were not disappeared from the political scene of Athenian democratic *ecclesia* (Hansen 1983, 207–8). They sustained their political influence among members of the *ecclesia*.

Regardless of geographical, cultural, and historical origin, permanent marketplace belongs to long-standing socioeconomic activity that establishes and sustains urbanism. In various traditions, a market is a periodic event on usually once a week basis at a public urban place that gathers people for trade and social interaction. The happenings of the market day take place at the location of what we understand the center or core of urban settlement. In most cases, marketplaces become the traditional destinations of going and visiting urban areas. Nonetheless, the urban settlement is not only reliance on the establishment of the permanent marketplace, but also the institutions of authority for laws and rules that protects, keeps, and sustains peaceful interactions among people as urban citizens. In democratic societies, such authorities have differentiated roles and responsibilities such as executive, legislative, and judicative functionality. These three bodies of governance work and exist to steward and defend urban statutes for freedom, social justice, human dignity, growth, and prosperity.

The necessity for clarity and determination of rules and orders underpins the establishment of urban statutes as categorical imperatives of conducts and interactions in urban areas in terms of laws, acts, regulations, and decrees. As an operational and referential system of urbanism, urban statutes include the ideas, principles, practices, and efforts of urbanity. In other words, urban governance is a politically organized system that protects, serves, and works for urban citizens based on urban statues. Nevertheless, the practice of urbanity needs an institution that works on a daily basis to steward and ensure its proper implementation by all citizens. Such an institution is in the role and function of monitoring, observing, controlling, and preventing violence against principles of urbanity; in modern urbanism, such an institution is known as the municipal police.

The existence of a permanent marketplace in various forms and manners seems to be the determinative factor for the establishment of urbanity and its settlement. The essence of the market lies in the necessity for a gathering of people for exchanges of foods, goods, and other commodities based on trust and fairness. The existence of the marketplace shows how important are exchanges and social interactions for urbanism. The existence of a marketplace builds the meeting point and relationship between the production and distribution of commodities. From this meeting point, new demands, services, and other related activities come about and grow in concert with the development of populations. Without marketplace, all human productions, as well as works, do not have any chance for mutual exchange that generates growth for quality and quantity.

The significance of marketplace as urban phenomenon lies in the happening of gathering that builds up and sustains a human relationship for developing products and services in its fine and delicate manner. Creativity and productivity of the land-based economy and human skill and knowledge do not have any improvable learning curve without the existence of the marketplace. Exchange in the manifold of form, matter, and way is the fundamental mechanism of urbanity so that interaction, production, and distribution are manageable within the framework of urban society as a whole. The nature of exchange comes into being in response to the necessity for interaction and communication. Exchange enables us to find more possibility based on interaction and communication. Barter is actually the form of economic exchange based on mutuality.

The necessity for exchange is on the ground of the human condition that man is human because of communication and interaction with others. Without exchange, man does not have any sense to live as a social being. Exchange ideas and experiences enable people to grow on the ground of mutual understanding. The importance of exchange is to sustain human condition with respect to the fact that being is for a man always being with others within the framework of mutuality. Regardless of geographical and historical circumstances, urban is the general concept of human settlement in the form of sociopolitically organized entity that lives and sustains their household within the system of community based on the exchange. The mechanism of exchange comes about on the ground of surplus in the agricultural economy. Interaction, collaboration, and specialization come into play for elaborating and developing human possibilities that manifest in manifold productions and services. All this is only possible because of the establishment of laws and orders.

There is no urban phenomenon without the exigency of law and order, more, and manner which constitute the possible condition of urbanity. Without the system of law and order, living with highly concentrated populations is potential in chaos; the system protects people from possible violence, abuse, and misdemeanor. Within the urban system, every citizen is responsible by law for keeping urbanity alive. In doing so, political engagement is part of urban citizenship so that urban society as a whole is active to nurture themselves as civil society. Disengagement and apathy in urban political processes give us an indication of unhealthy urbanity.

Historically, the establishment of the urban site is not simply for conquering natural land because without being able to domesticate its resources, the settlement does not come into being for its sustainable future. Rather, the necessity for making an urban enclosure lies in the ground of freedom from natural bondage under the concept of cultivation toward civilization. Making urban settlement is historically an evolution of land and human cultivation. Essentially, urbanism is the act of making up with the natural environment for social living. By making up, the topography of the site becomes the guiding principle for building layout, drainage, sewage, communication, and accessibility. Historically, building an urban settlement is characterized by the establishment of sociopolitical authority. The establishment of authority is evidently necessary for instituting law and order that protects peace and justice.

The foundation of urbanism in democratic society lies in its productive economy that generates growth and prosperity based on equal opportunity for jobs and social justice for its services to citizens. Democratic governmentality of the municipality is to maintain and sustain the livability and sustainability of urbanism. Knowledge and technique of production are determinant factors of the system of urbanism that manage and control urban production in regard to the quality of life of its urban citizens and environment.

Since dwelling for a human being is actually living together with others, the urban settlement is nothing but the phenomenon of socioeconomic and political engagement and commitment of a group of people living in a permanent location. Under the notion of society, human beings are members of a creatively living system with the common awareness that grows and develops their productive capacity by the act of making. The membership of professional or interest-related associations has been known as an integrated part of urban society for their work, live, and play since ancient times. The establishment of various social groups for work, live, and play had played an important role in urban society since ancient times. Besides the need of social interactions and the updating of agreements for the rules of conduct, the intention of the establishment of such groups and clubs is their common interest in the pursuit of highly specialized skills and businesses.

In contrast to the village, the urban settlement is not a communal polity but a state polity that establishes and sustains highly sophisticated society and civility. Society is distinguished from the community because of its capacity to manage and control their resources for their social, economic, cultural, and political cultivation. A community is not urged for productivity that expands and develops their knowledge, skills, and domains for new opportunities. Rather, a community is to develop its capacity to maintain and sustain a homeostasis within their production system (Trojan 1984, 23) so that expansion and development are limited in a way of refinement and elaboration, instead of exploration and accumulation.

Nevertheless, society as a sociological concept is not simply a collectively established social system of living. Rather, under the notion of society, human beings are interwoven within a rationally interrelated system of exchanges of information, commodities, and services. Consequently, society as a whole incorporates manifold human activities for cultivating their skills and knowledge in dealing with development based on intensive training and learning. The sense of society lies in the diversity of activities, associations, and specialties within the framework of single purpose for beautiful living. The importance of diversity for the urban system lies in learning from each other based on their differences for something good and healthy. Indeed, diversity engenders a variety of ideas and approaches as well. However, diversity is necessary for trade and state because it creates and brings about new demands and challenges for new things, improvements, elaborations, and collaborations. Historically speaking, great cities of the globe such as New York, London, and Paris are not only racially, socially, and politically diverse, but also culturally and economically various.

In order to maintain and sustain their welfare, urban society as a whole is provided with the system of law and order that protects and keeps the commonality based on rights and justice. The emergence of law as the foundation of rules of conduct and social interaction is not simply by the reason of communal living. Rather, the necessity for the law is to ensure the permanency of commitment and involvement of people as the members of society. Laws are more than customary practices in the community-based settlement. The sense of law lies in its institutionally established system of rules of conduct in keeping and sustaining urbanity.

In its daily life, the center manifests as the meeting place where the establishment of the identity of arrival and departure is signified architecturally. Institutionally, the phenomenon of the center is identified with the seat of authority for laws and orders and the meeting point of urban citizens for sociopolitical and economic activities. Besides the establishment of an authority with various institutions for socioeconomic and political activities, the emergence of labor societies or guilds belongs to the phenomenon of urbanism (Saunders 1995, 18).

The transformation from village to urban settlement is not simply by the reason of population growth. Rather, the alteration is on the ground of the necessity for permanent authority that maintains, protects, and sustains the interactions and exchanges among people. Indeed, urban life is more than just economic subsistence; it is somewhat productive with a surplus economy which is based on agriculture, industry, and commerce. The leading activities in production or in service generate other economic sectors and services that bring about the interrelation and its mutuality of employment within the system of urban welfare. In most cases, the dominant commodity or service establishes urban economic identity, in terms of local competitive advantage. Of course, urban identity is not only about economic strength, but also it is provided with everything extraordinary such as; cultural institutions, annual events, history, and architecture, the town and city can proffer that distinguishes them from others.

Even though human community lives in the natural environment with respect to their physiological property and identity, human beings are creative beings that make up their environment. The natural environment is never ready to live on site for human beings. Making place is the way of human beings dealing with the natural site. Accordingly, the natural site is always the subject of making up for human beings that enable them to establish their collective living. The necessity for a collective dwelling is on the ground of their essence that is already gifted as members of the community. Human beings are essentially social not because of morality. Rather, they are social because only in this way they naturally survive for liberating themselves from natural bondage.

The division of labor is actually not by the reason of choice and interest, but it is essentially the nature of the work itself with respect to the growing demand for quality and quantity. Since every work nearly demands certain knowledge and skill, the division of labor is nothing but the response to handle trade appropriately which is provided with special and intensive learning and training. Therefore, urbanism is characterized by diverse learning and training institutions which belong to the urban culture. The purpose is to establish and develop a sustainable system that keeps up the quality of urban life for a sustainable future.

The urban dwelling is actually society living in which fair competition and professionalism constitute and sustain its urbanity. Of course, there is always room for social solidarity and mutual cooperation. However, under the notion of urban mutuality and solidarity, urban dwellers are necessary to participate and contribute their part in various specific functions and roles in the making of productions, maintaining regulations and statutory matters, delivering of services, training, and educations. These various urban activities are succinctly organized into urban institutions which include public services, commercial, residential, rehabilitation, cultural educational uses, maintenance facilities, and supports. The existence of such institutions reflects and signifies the diversity of urban care for urbanity. Such urban institutions do not only nurture manifold human possibilities but also enrich and enhance human potentialities in a sustainable way.

However, based on its main purpose, urban institutions and land use can be comprised of three categories: work, live, and play. Residential use is the most dominant area of urban area. It includes other facilities and infrastructure that support its well-functioning system as home. All activities related to the category of work are to deliver products and services. Recreational and open space belongs to essential use that upholds and sustains the other uses. Under the notion of play, we understand the site where urban citizens find their place for refuge and relaxation from work and live. To be intact with the natural environment is one important aspect of this place.

Even though wage is not the only way to value works, the compensation of the labor in delivering products and services becomes essential for the urban economy. Remuneration with currency is not only a transferable compensation of the work and labor, but also an urbanite way of reward and acknowledgment for human skills and endeavors. On the other hand, living is to comprise all activities for taking care of our well-being such as education, recreation, leisure, and domestic. Living and working are the main components of urban life; both activities are constituted by the necessity of human beings—either as an individual or social beings—for their nourishment and creative endeavors. Balancing of the quality of living and the productivity of working belongs to the essential enterprise of urbanity. Urban productivity is necessarily supported by the basic condition of healthy nourishment in which the surplus of agriculture plays its significant part. Besides human resources, natural resources in the form of foods, goods, and other materials become the important input of the urban system of production. Any successful urban vicinity or place is mostly supported by steady and sustainable resources within its reasonable reach.

Concluding Notes

Town or city is not simply the built environment where we live, play, and work. Rather, it is the world where we establish, develop, and cultivate our civility and civil society. Traditionally, we call a place as town and city because of its concentration of populations, its compactness of settlement, and a well-consolidated institution for economic, sociocultural, and political activities. Historically, the necessity for dwelling in proximity and gathering underlies the embryonic establishment of urban settlement institution provided that agriculture and exchange sustain its domestication. The permanent marketplace must have taken place at the center of the urban settlement. Nevertheless, only stable sociopolitical institution keeps up the economic sustainability of human settlement in development. Without such sustainability, it is difficult to grasp a compact settlement area as a living town. The economic sustainability and stability of political institution guard the existence and growth of urban settlement.

Historically speaking, town or city is always associated with the phenomena of cultivation of human society with an architecturally concentrated settlement. The concentration of dwelling makes intensive interaction and communication that constitute mutuality for exchange and organization for production and cultivation. Based on this compactness, the intensively interacting and exchanging processes take place in an efficient manner. We speak of cultivation in the context of human achievements for refining our understanding of being through actions, thoughts, and feelings. The works of art in various forms and media become the most important things of urban production. Without the works of art, urbanity does not perform its tangible outcome.

By the establishment of town or city, we cultivate everything proper for our being as urban society as a whole that lives and grows together. Cultivating here is not only practically improving and refining something with labor and work, but also ethically that with others based on respect and dignity. Provided that the sense of urban society as a whole lies in its synergic diversity of people for economic and cultural systems of production and service, cooperation based on fair share and responsibility is necessarily the foundation of urban society as a whole. Members of urban society as a whole should have their own share and responsibility under the principles of mutually sustaining production, service, and regeneration, otherwise urbanity would gradually diminish and fall into ashes.

There is nothing more striking for urbanism than the relationship between a public sphere and civil society. The presence of the public realm is ontologically necessary; without this, urbanism does not fulfill its commitment to civility and civil society. Because both are considered as the sense and essence of urbanism. Without both concepts, urbanism does not have the ontologically constructing reality of the urban life-world as a political community.

References

Andranovich, Gregory, and Gerry Riposa. 1993. *Doing Urban Research.* New York: Sage.

Arendt, Hannah. 1958/2013. *Human Condition.* Chicago: University of Chicago Press.

Aristotle. 1997. *Politics, Volume IV.* Edited by Richard Kraut. Oxford: Oxford University Press.

Aristotle, and Richard McKeon. 1992. *Introduction to Aristotle.* New York: Modern Library.

Banhart, Robert K., and Sol Steinmetz. 1988. *The Barnhart Dictionary of Etymology.* Hackensack, NJ: H. W. Wilson.

Binne, Jon, Julian Holloway, Steve Millington, and Craig Young. 2006. *Cosmopolitan Urbanism.* London and New York: Routledge.

Blackwood, William. 1822. "Blackwood Magazine." *Edinburg Magazine,* 346.

Bogucki, Peter. 1999. *The Origins of Human Society.* London: Blackwell.

Bowman, Alan K., Peter Garnsey, and Dominic Rathbone. 2000. *Cambridge Ancient History,* vol. 11. Cambridge, UK: Cambridge University Press.

Branigan, Keith. 2002. *Urbanism in the Aegean Bronze Age.* New York: Sheffield Academic Press.

Bravi, Alessadra. 2015. "The Art of Late Antiquity: A Contextual Approach." In *A Companion to Roman Art,* edited by Barbara Borg, 130–50. Malden: Wiley.

Calthorpe, Peter, and William B. Fulton. 2001. *The Regional City Planning for the End of Sprawl.* Washington, DC: Island.

Catanese, Anthony James, and James C. Snyder. 1979. *Introduction to Urban Planning.* New York: McGraw-Hill.

Cohoone, Lawrence. 2002. *Civil Society: The Conservative Meaning of Liberal Politics.* Malden: Wiley-Blackwell.

Dillon, Sheila. 2006. *Ancient Greek Portrait Sculpture: Contexts, Subjects, and Styles.* Cambridge: Cambridge University Press.

Ellin, Nan. 2013. *Good Urbanism: Six Steps to Creating Prosperous Places.* Washington, DC: Island Press.

Foucault, Michel. 1985. *History of Sexuality, Volume 2: The Use of Pleasure*. Translated by Robert Hurley. New York: Vintage.

Fulton, William B. 1996. *The New Urbanism: Hope or Hype for American Communities?* Cambridge, MA: Lincoln Institute of Land Policy.

Gates, Charles. 2003. *Ancient Cities: The Archaeology of Urban Life in the Ancient Near East and Egypt, Greece and Rome*. New York and London: Routledge.

Geddes, Patrick. 1915. *Cities in Evolution*. London: William & Norgate.

Gidden, Anthony. 1990. *The Consequences of Modernity*. Stanford: Stanford University Press.

Habermas, Jurgen. 1989. *The Structural Transformation of the Public Sphere: An Inquiry into a Category of Bourgeois Society*. Translated by Thomas Burger and Frederick Lawrence. Cambridge, MA: MIT Press.

Hansen, Mogens Herman. 1983. *The Athenian Ecclesia: A Collection of Articles 1976–1983*. Copenhagen: Museum Tusculanum Press.

———. 1991. *The Athenian Democracy in the Age of Demosthenes: Structure, Principles, and Ideology*. Translated by J. A. Crook. Norman: Oklahoma University Press.

———. 1997. *The Polis as an Urban Centre and as a Political Community*. Copenhagen: Royal Danish Academy.

Harris, Edward M., David M. Lewis, and Mark Woolner. 2015. *The Ancient Greek Economy: Markets, Households and City-States*. Cambridge, UK: Cambridge University Press.

Heidegger, Martin. 1927/2010. *Being and Time*. Edited by Dennis J. Schmidt. Translated by Joan Stambaugh. Albany, NY: SUNY Press.

Jacobs, Jane. 1961. *The Death and Life of the Great American Cities*. New York: Random House.

Katz, Peter, Vincent Scully, and Todd W. Bressi. 1994. *The New Urbanism: Toward an Architecture of Community*. New York: McGraw-Hill.

Kelbough, Douglas, and Kit Krankel McCullough. 2008. *Writing Urbanism: An ACSA Publication*. London and New York: Routledge.

Kipfer, Barbara Ann. 2000. *Encyclopedic Dictionary of Archaeology*. New York: Springer.

Kostof, Spiro. 1991. *The City Shaped: Urban Patterns and Meanings Through History*. London: Thames & Hudson.

Kraut, Richard, and Steven Skultety. 2005. *Aristotle's Politics: Critical Essays*. Lanham, New York, and Toronto: Rowman & Littlefield.

Lynch, Kevin. 1981. *A Theory of Good City Form*. Cambridge: MIT Press.

McClain, James L., and Osamu Wakita. 1999. *Osaka: The Merchants' Capital of Early Modern*. Ithaca, NY: Cornell University Press.

Mcclelland, J.S. 2015. *A History of Political Thought*. London and New York: Routledge.

McIntosh, Jane, and Clint Twist. 2001. *Civilizations: Ten Thousand Years of Ancient History*. London: BBC Worldwide.

Miles, Malcolm, Tim Hall, and Iain Borden. 2004. *The City Cultures Reader*. New York: Routledge.

Mumford, Lewis. 1961. *The City in History: Its Origins, Its Transformations and Its Prospects*. New York: Harcourt, Brace and World.

Passell, Aaron. 2013. *Building the New Urbanism: Places, Professions, and Profits in the American Metropolitan Landscape*. London and New York: Routledge.

Pomeroy, Sarah B., Stanley M. Burnstein, and Walter Donlan. 1998. *Ancient Greece: A Political, Social, and Cultural History*. New York: Oxford University Press.

Rae, Douglas W. 2005. *Urbanism and Its End*. New Haven: Yale University Press.

Ramirez-Lovering, Diego. 2008. *Opportunistic Urbanism*. Melbourne: MIT Press.

Roberts, Lance W., Rodney A. Clifton, and Barry Ferguson. 2005. *Recent Social Trends in Canada, 1960–2000, Comparative Charting of Social Change*. Montreal: McGill-Queen University Press.

Sanders, William T., and Barbara J. Price. 1968. *Mesoamerica: The Evolution of a Civilization*. New York: Random House.

Saunders, Peter. 1995. *Social Theory and the Urban Question*. New York and London: Routledge.

Scott, Alen J., and Michael Storper. 2015. "The Nature of the Cities: The Scope and Limit of Urban Theory." *International Journal of Urban and Regional Research* 39 (1): 1–15.

Sennet, Richard. 1977. *The Fall of Public Man*. New York: Alfred A. Knopf.

Short, John Rennie. 2014. *Urban Theory: A Critical Assessment*. London: Palgrave.

Simmel, Georg. 1950. "The Metropolis and Mental Life." In *The Sociology of Georg Simmel*, edited by D. Weinstein, translated by Kurt Wolff, 409–24. New York: Free Press.

Steinberg, Philip E., and Rob Shield. 2008. *What Is a City? Rethinking the Urban After Hurricane Katrina*. Atlanta: University of Georgia Press.

Talen, Emily. 2013. *Charter of the New Urbanism*. New York: McGraw-Hill.

Tarling, Nicholas. 1999. *The Cambridge History of Southeast Asia*. Cambridge: Cambridge University Press.

Thomsen, Rudi. 1964. *Eisphora: A Study of Direct Taxation in Ancient Athens*. Oslo: Gyldendal.

Trojan, Przemyslaw. 1984. *Ecosystem Homeostasis*. Berlin and New York: Springer.

Vernant, Jean-Piere. 1983. *Myth and Thought Among the Greeks.* London: Routledge & Kegan Paul.

Waldheim, Charles. 2012. *The Landscape Urbanism Reader.* New York: Princeton Architectural Press.

Wirth, Louis. 1938. "Urbanism as Way of Life." *American Journal of Sociology* 44 (1): 1–25.

Zijderveld, Anton C. 1998. *A Theory of Urbanity: The Economic and Civic Culture of Cities.* Piscataway, NJ: Transaction.

3

Lesson Learned from the Ancient Greek Polis

Urbanism in Ancient Greek Society

What is urbanism in ancient democratic Athens under Pericles? There are the studies on the architecture of the ancient Greek *polis* that have been explored by several scholars (Martin-McAuliffe and Millette 2017; Preston and Owen 2009; Hakim 2014). However, there is still one of the many aspects of the ancient Greek urbanism that has been less divulged and unfolded especially in relation to the Aristotle's politics, namely urban democratic culture and governmentality. Meanwhile urban culture in urban studies is mostly associated with media, urban literature, cinema, public arts, sports, and entertainments (Cowan and Steward 2007; Jencks 2004; LeGates and Stout 2003, 87–136), the sense and essence of urbanism in democratic societies still need exploring profoundly, especially concerning the notions of culturally urbanized practices and products. Such concepts are considered resourceful for the reason why people as a sociopolitically organized community build their urban settlements for their homeworld. In doing so, the notion of home is an integrated system of place for living, working, and recreation. However, the integration of human life and their place cannot ignore the spiritual necessity for the

© The Author(s) 2020
B. Wiryomartono, *Livability and Sustainability of Urbanism*,
https://doi.org/10.1007/978-981-13-8972-6_3

sense of collective habitation with freedom and dignity; this is what is experienced as civility.

In order to explore the relationship between civility and urbanism, this study argues that the ancient Greek *polis* is one of its solid foundations because of its Athens' democratic governmentality in circa the fifth BC. Notwithstanding its technological and societal advancements and improvements throughout the ages, the intentionality of urbanism remains the same. From *polis* to today city, the intentionality directs all endeavors toward the highest common interest and the best. Democracy and Solon's code of laws had been the structures of the Athenian political culture; Pisistratus governed Athens according to laws as described by Aristotle (*Politics* 16.8 cf. 14.1, 16.2) in terms of *demotikos* and *philantropos* (Ober 2011, 356).

Historically speaking, the ancient Greek democratic experiences (500 B.C.–200 B.C.) have put the urban political culture together in the concept of the *polis*. Accordingly, *polis* is defined as the city-states in ancient Greek times from circa sixth century to 338 B.C. Aristotle (384–322 B.C.) wrote his work: politics between 335 and 323 B.C. (Miller 2014). In this context, there is nothing essential for urbanism but the decorum of humanity based on law (*nomos*) and order (*taxis*). In its time frame, it is about the *polis* in the time before the Charonian war, which was the milestone for the decline of city-states as autonomous polis (Hansen 1993, 21).

Urbanism and Aristotle's *Polis*

Aristotle's *Politics* is one of the most ancient written sources of urbanism available on the globe. Aristotle's concept of *polis* is considered pivotal for the theoretical foundation and framework of this study. In Aristotle's *Physics* II.3, city-state or *polis* can be seen as the world in an architecturally complex thing comprising of the four *aitia* (Kalimitziz 2000, 107–8); the *aitia* consist of matter (*hyle*), form (*morphe and eidos*), potentiality (*dynamis*), and activity (*energy*). Accordingly, *polis* is a system of human settlement with certain structure and form that contains various elements and works in a certain location—*topos*. However, *polis* is not any human settlement but a specific world with an autonomous political culture.

As noted by Aristotle in *Politics* Book V, the existence of *polis* is to hold its assembly and constitution away from *stasis*, *which* means in crises and politically divided (Price 2001, 31–2). As the world, *telos*, of the city-state is about the integrity between the life-world and its societies based on *nomos*—laws, norms, codes, and manners—the only danger for *polis* is its *stasis* or disintegration. *Stasis* includes internal conflicts, political disturbances, or contestations that put the constitution of the *polis* in peril. In such circumstances, the existence of *polis* is on the brink of danger.

In ancient Greek times, the decorum of humanity is inscribed and circumscribed with the concept of the *polis*; this concept is to distinguish its happenings from barbarity (Fitzgerald 2007, 109–6). In the context of the *polis*, Aristotle signifies this decorum in terms of *arete*; the highest good of human manners and conducts, which is inevitably important for societal well-being (Garver 1994, 110–14). It is more than moral virtue because the sense of *arete* lies in its functional concept that citizens are able to do their best in a way of being functional and being respectful in their society. Succinctly speaking, *arete* is about honor based on individual merit and ability (Jaeger 1939/1986, 8). In the Classical and Hellenistic *polis*, the existence of court institution, *boule*, is essential representing an independent political body that upholds the significance of justice in maintaining and sustaining human decorum. Moreover, the establishment of a court of law in *polis* implies the importance of ethical principles for civility (Moris and Powell 2010).

Civility in ancient Greek times is socioculturally conditioned by the *polis* (Barker 1948; Pagden 2002, 40). The common translation of the concept into English is 'city-state' that means an urban settlement with an independent government. The Latin *civitas* is a translation of the Greek *politiea* for the notion of citizenship of a political community of urban settlement. The concept of civility is in the Latin word *cives* meaning citizen that is similar to the Greek *polites*. Accordingly, civility has something to do with the condition for 'being an urban citizen', in reference to their proper behaviors and good manners. In the context of *polis*, civility is considered essential and existential in terms of ethical norms, codes, and customs. However, civility is a product of city life, but the life in the regulated state, *civitas* (Fitzgerald 1901, 283). In other words, *polis* is the ancient Greek world in a sociopolitically ordered system based on the supremacy of law with a democratic assembly of representatives of the citizens.

Urbanism and Education

The relationship of urbanism and education cannot ignore the necessity of sustainability in terms of economy and culture. This bond is about the regeneration and growth of urban settlement and populations for betterment. However, the sense of urban economic and cultural growth is actually experienced by people in the public realm (Orr 2004; Julier 2000; Heller 1997). The notion of civility of *polis* comes into the light as the respectful and well-mannered daily life-world that is understood by Aristotle's Politics as *arete*. In ancient Greek times, *polis* has been always the site of the Greek culture. It is the place where Greek language, customs, and behaviors are properly exercised within the framework of *nomos*—rules of law. Those who do not speak and behave the Greek language and manners properly are called *barbaros* (Harris 2004, 51). The pleasure of the lively experienced interactions in the public realms is associated with the sensually uplifted sense based on mutual respect and sociability; Aristotle conceives this pleasure in terms of *schole*. Accordingly, *schole* is the aesthetic experience that cultivates the mind with new things and wisdom.

In *Politics* (Book VII 1.4. 1333 16–30), Aristotle describes two kinds of experience, which are *schole* and *diagoge*. *Diagoge* is distinguished from *schole* with the joy of doing something on routine and regular basis; *diagoge* cultivates the mind and body with the skills of habitation. Meanwhile, *schole* is the delight of doing things that cultivate the mind with relaxation and entertainment. According to Aristotle, only in *schole* man is his own master, not serving anybody but his own wish (Eriksen 1876/1976, 49). *Schole* is often translated into English language with the words: play, leisure, and amusement, while *aschole* is associated with routine and boring activities. To certain extent, the Greek concepts of *schole* and *diagoge* are hardly translatable. Aristotle puts *diagoge* in the context of *paideia*—education, while *schole* is associated with the restoration to health and recreation (Nagel 2002, 48–9).

Regarding the cultivation of mind and body, the aesthetic experience in terms of *schole* and *diagoge* is considered neither hedonistic nor frivolous pleasure. Moderately speaking, both concepts are the mindful delight of engagement and being fully integrated with the presence. The basic structure of this engagement is made possible through mindful practice and exercise *melete*, which means being aware with care and attention; Plato to Diotima, *Symposium* 208 A-B, mentions *melete* as the way toward eternity. In the context of *melete*, activities are experienced as the engagements of mind and body with spiritual joy such as listening to empathetic music, reading of resourceful texts, and watching an interesting game. In contrast to *techne*, *melete* has not something to do with production. Plato in Phaedo points out the sense of *melete* in the relation to 'cultivation of' and 'preparation for'; the word *melete* refers to the verb *meletan* meaning occurs regularly. Cebes responses to Socrates in Plato's dialogue concerning the initial depiction of philosophy as *melete thanatou*—cultivation for death (Jordan 1993, 88–9). In contrast to *techne*, Heidegger understands the concept, *melete* is care that is capable to bring back the gift of presence from its fleeting nature; while *melete* preserve the engagement with the presence of being, techne is the capability for revealing and disclosing beings from absence to presence (Heidegger, *Nietzsche Volume I & II* 1991, 164–5; Krell 1988, 164–6).

From the perspective of Aristotle's *polis*, being in the *polis* is a human condition for their societal well-being or good living (Barker 1958/1998, xii); it is practically made possible by law that engenders amenable sociability based on the constitution, *nomos*. Moreover, *polis* is a well-organized built environment with various domains and things that matter for humankind as political being. Thus, *polis* is not any settlement but a political institution of societal habitation and economic community. For Aristotle, *polis* is a system of geographical location (*topos*) and community (*koinonia*), which are organized and established with law and order (*nomos*). For Aristotle, the intention of polis is to nurture and cultivate humankind toward the sense of *eudaimonia* (*Nicomachean Ethics* 1095 an 18–20). Accordingly, *eudaimonia* is considered the concept for the state of well-being and flourishing. Eudaimonia is often translated as happiness but for Aristotle, the concept is more than that. Aristotle understands the concept in terms of the human fulfillment as sociopolitical being in society.

In Aristotle's *Nicomachean Ethics* 1177b 9–34, *eudaimonia* is the state of being that people feel good about themselves because of the fulfillment of their functional occupation and of the ability to do everything according to the greatest things in their potentialities (Gonzales 2006, 136; Haworth 2004, 52–3).

Ethically speaking, *eudaimonia* is the state that one is able to reveal oneself, -as fully functional and professional person-, for doing the excellence in the society. Aristotle speaks of the highest good in ethical terms of eudemonia in *Nicomachean Ethics* Book I 2. In the *polis*, the necessity of the law leads people to engender the greatest good or well-being. Aristotle's *Politics* I.1 1252a 1–7 points out the purposeful fullness of the *polis* is civility that everyone does everything for the sake of the highest good by law.

Architecturally speaking, the *polis* is the *topos* of human decorum because it is provided with the marketplace *agora*. In doing so, fair and respectful socioeconomic activities are able to be practiced in a public realm. The importance of *agora* is not only to provide *polis* and its citizens—*politea*—with the sense of place but also with the public institutions of exchanges and social interactions based on the common interests and fairness. The public realm of *agora* guards and sustains the civility according to certain ethical codes and principles—*nomos*. By and large, *agora* is the core of urban livability of the Greek culture and tradition. In other words, the Greek *agora* is the platform of *polis* that enables people to demonstrate the excellence for the functional completion of urbanism. Every human effort and endeavor in the Greek culture are necessary to have this functional completion—*telos*—which is commonly translated as a finality, end, and perfection.

In *Physics* II.2. 194a 29–30, Aristotle describes *telos* in relation to the presence of things for the sake of they are. Accordingly, *telos* is not simply to say about the fulfillment of a product, an endeavor, of effort. Rather, *telos* is to say that things are perfectly in function according to its assignment and its own sake, for which it is designed and constructed. In other words, *telos* is to say that things are fully functional in the world with specific task and designation. However, Aristotle, in *Nicomachean Ethics*, does not say that death is *telos* but the merit of life in terms of the best he or she can do. In this line of thinking, one's *telos* is his or her well-known works or deeds in according to his/her capability and service to humanity. Aristotle (*Physics* II.7.198 25–26) points out about *telos* that 'What the thing is, what the thing is for, are one and the same' (Gonzales 2006, 131–3).

Moreover, *polis is* a human system that is constructed with a functionally originating principle, *arche*; Aristotle in Physics B1 conceives *arche* as the originating source of the circle of gratitude and responsibility—*aitia*—that engenders motions, transformations, and processes in the nature. In Aristotelian *physis*, nature and human system are characterized by something in common for its origination and structural principle because both are necessary in compliance with the natural laws such as gravitation and thermodynamics. Human system needs *arche* for composition, construction, and operation so that the artificial things are able to fit in the natural environment. The way of how humankind deals with the productions of things is understood by the Greek as *poesis*; in the concept of *poesis*, the Greek does not only know how to make it but also how to use and take care for it. However, *techne* is a kind of *poesis* because of *arche*. The highest effort and endeavor of making is *techne* that leads forth in terms of improvement, betterment, and refinement, such as from the raw to the cooked, from the simple to the sophisticated as well as from the coarse to the fine. Founding and developing urban place are the *poesis* of human settlement on certain location, -*topos*- that enables the life-world to exist and grow. The development of *polis* is not simply determined by the number of populations or by the complexity of its buildings and roads. Rather, *polis* is the domain of a nation in terms of *koinonia*—community or association of societies—with a constitutional governmentality—*politeia*. Nevertheless, the sustainability of *polis* cannot disregard the existence of an educational institution and its tradition under the notion of *paideia*.

Paideia is essentially the source and foundation of the ancient Greek culture and tradition. Based on his longstanding historical study of ancient Greek culture, the scholar of Greek civilization Werner Jaeger (1939) concludes that the existence of *paideia* sustains and develops the *polis* generation to generation. Historically speaking, *paideia* as a concept is which has been deeply rooted in Greek aristocracy circa the fifth century B.C. The very idea of *paideia* is the concept of respect, pay homage to gods, respect to parents, and respect for strangers. Based on the concept of respect, the

ancient Greeks develop and sustain the cultivation of aristocratic manners and behaviors. The ancient Greek aristocrats enjoy their nobility
status because of their *kalos kagathos* meaning decency through speech
and action. The ultimate achievement of this nobleness is *arete*, which is
untranslatable because of its manifold meanings. The concept of the *arete*
is a synthesis of the sense of confidence and gallantry with the courage of
warrior. Within the framework of *paideia*, the ancient Greeks conceived
the quality of humankind for their outstanding skills, surpassing physical
strength, and inner determination as the basic condition of leadership. The
city-state culture had been the educational center of nobility of the ancient
aristocracy, which was characterized by, according to Homer's Iliad, two
ethical concepts: *aidos* (being ashamed of disgraceful action) and *nemesis*
(retributive justice, righteous indignation) (Jaeger 1939/1986, 3–34).

Urbanism and Gathering Place

Since prehistoric times, the need for being a part of the crowd has been
manifesting in the phenomenon of the marketplace. The urban crowd
of the ancient Babylon came together to *bazaar*, the Greek to *agora* to
the European Middle Age to farm marketplace and modern society to
shopping mall. The marketplace is more than just a place of trade. Rather,
the marketplace is the center of gatherings and exchanges in the broadest
sense of the word that socioeconomically establishes and develops urban
community. Historically speaking, the word *agora* was to address a place
of political gathering, a place where urban citizens came together to make
deliberations of their political affairs (Moller 2000, 71). In Homeric epics,
agora was often associated with the assembly of urban citizens, the council
of elders, and the advisory boards of *polis*.

The joy and cheerfulness of casual gatherings and exchanges have been
characterizing and cultivating humankind for the sense of sociopolitical
being since ancient times. The need for mere economic exchange of commodities had been overcome and enriched by the ancient Greeks with the
manifold needs of sociopolitical being so that *agora* became the center of
cultural development for the ancient Greek society. Accordingly, the *agora*

of the democratic Athens was the hearth of freedom and civil society. Aristotle (*Politics* 1331a 30–6) underscores the significance of free *agora*, which is provided with *gymnasia* for physical exercise, public education, and discourse on sociopolitical and economic affairs. Accordingly, *agora* for Aristotle is more than just the marketplace (Aristotle and Kraut, *Politics* 1997, 122).

Indeed, the marketplace is not only a place for the purchasing and sale of provisions and livestock and other commodities. Rather, marketplace is the gathering place of people where people are able to exercise, cultivate, and sustain their freedom, trust, fairness, and mutual respect regularly; all of these virtues are considered essential and necessary for democratic urbanism. They are essential and necessary because the virtues are constituent for human decency prior to any law and regulation. The happenings of the market provide people with the opportunity for free communications and interactions. Historically speaking, marketplace (*agora*) was always the place where the Greek virtues of fairness (*epieikeia*), equality before the law (*isonomia*), and good manner (*kalokagathia*) were practiced and developed on a regular basis. Aristotle recognizes the interpersonal virtues among citizens in terms of *philia*; the concept is to describe the friendly relations among people that lead toward social nourishment and self-actualization for the excellence. Because of this social atmosphere, the Athenian democratic *polis* existed for the sake of noble actions (Voegelin and Germino 2000, 393).

Urbanism and Urban Intentionality

One important question concerning urbanism is why is humankind directed into the excellence of endeavor and behavior? In the ancient Greek life-world, why were they collectively directed and drawn to do for the excellence in terms of *arete*? This is the utmost virtue of all ancient Greek efforts and endeavors as well as behaviors in the context of *polis*. The highest achievement of the ancient Greeks cannot ignore the significance of *arete* that has spiritually moved and driven to do their best to enjoy their spiritual fulfillment—*eudaimonia*. The *arete* is considered the essential urbanism of *polis* while *eudaimonia* is the intentionality of collectively and subconsciously driving forces, efforts, and endeavors of the ancient Greek urban citizens for urbanity.

Urbanity is the spirit of urbanism that is developed and cultivated through the sociocultural and economic interactions. Urbanity is not simply the outcome of urban development and cultivation. Rather, urbanity is an open-ended system and process of urban activities that brings about the sense of habitation and the freedom of humankind from fear and alienation. The sense of habitation includes the experience of respectful socialization and amenable bond of urban citizens. Urbanity was experienced by the ancient Greeks in their gatherings in public realm (*koinon*) (Gernet 1981, 319); such experience had been part of urban life in *agora*, amphitheater, and gymnasiums. Generally speaking, all activities of the ancient Greeks were connected and affiliated with the common domain (*koinon* or *koinonia*) (Taylor and Vlassopoulos, *Communities and Networks in the Ancient Greek World* 2015).

Being *urbane* is another word for urbanity that depicts the state of affairs and conditions of politeness and refinement. As matter of fact, urbanity of the ancient Greek *polis* was not simply provided by the culturally developed urban citizens. Rather, this urbanity was factually on the architecturally built environment with conducive public realm, which was integrated with culturally nurturing and cultivating activities for mind, body, and spirit. As an architecturally constructed thing, the ancient *polis* under democratic government was by design a well-organized place with a politically established system and compact form; all areas for collective activities in the city were connected with the common domain called *koinon* (Schmitt-Pantel 1991). Aristotle (1260b 41) speaks of the *polis* as *topos* as the group of domains with certain categorical organizations and functional institutions. Accordingly, *topos* is not simply a location or a site, but a domain of the sociocultural, economic, and political activities, which is necessary to incorporate a continuous network of public realm. The intentionality of the ancient Greek *Athenian polis* as *topos* was to achieve the *arete* and to constitute the *politeia* for its purposeful wholeness, *telos* so that all citizens were able to experience *eudemonia*.

Urbanism and the City

What is the relationship of urbanism and the city? The ancient Greeks built and developed their *polis* was not simply to have aesthetically appealing buildings. Rather, they erected the built environment to experience the network of communal areas so that various associations and interest groups were able to be managed into a viable system of political engagement and socioeconomic web of all populations; the slaves and metics were not passive objects of exclusion and exploitation but the active historical agents that established their own diverse webs and associations (Taylor 2015, 29). The relationship between the populations and their city was established a sociopolitical bond called *koinon*. Accordingly, the design of the city was to incorporate the communities and associations into the network of public domains. The web of public realms became the generator of livability in the ancient *polis* with its symposiums, gymnasium, agora, and Acropolis. During special occasions of the *polis* in the fourth century B.C., the religious rituals or cultic practices were participated by all populations, regardless their status either citizen or slave (Taylor 2015, 43).

The question concerning urbanism is inevitable to the intentionality of humanity for transcending from biological condition to spiritual being. The architecture of the city has historically demonstrated how this intentionality to come into being evident. Doxiadis' studies on the architecture of the ancient Greek cities reveal that the chiefly principle of the Greek spatial planning is the visual perception so that an observer, from a certain location, is able to have panoramic scenes of the livability of public realms and buildings (Doxiadis 1964). Accordingly, the layout of buildings, its proportions, and compositions are organized with optical and geometrical preferences. To what extent are the findings of Doxiadis confirmed by archaeological evidences is still an open question. As matter of fact, some sites of the ancient Greek city show the layout of buildings with potential three-dimensional quality; most of buildings are not juxtaposed within grid arrangement but in a unique position so that one is able to experience the spatial effect, which is not frontal but visually multifaceted.

The aesthetics of public realm in Aristotle's *polis* is about the *topos* in terms of the site and livable built environment of the public realm. Accordingly, the ancient Greek public places as the main part of the *polis* were more than just areas of social interactions. Rather, the outdoor living in the public realm was the ancient Greek *polis'* way of life (Hooper 1978, 8); in other words, the ancient Greek public realms were their life-worlds, in which the socioeconomic and political engagements of all populations took place as a culturally integrated system for the cultivation of *politeia*. Accordingly, the architecture of the ancient Greek city was indivisible from the democratic establishment of this *politeia* that upheld civility, *arete*. Unsurprisingly, the upshot of this establishment was a remarkable display of monumental urban buildings (Thomas 1993, 15). Then, the ancient Greek *polis* was not only notable for its grandiose edifices and public places but also noteworthy of its livable scenes and visually appealing panoramas from every spot of its urbanized areas.

Architecturally speaking, the public realms of the ancient Greek *polis* was best described with Aristotle's *topos* or with *chora* of Plato's Timaeus; Kevin Lynch describes the quality of such good urban form that comprises the manifold properties and dimensions of urbanism such as vitality, sense, fit, access, control, efficiency, and justice in the context of social uses (Lynch 1981, 118–9). As matter of fact, such culturally urbanized qualities cannot ignore the lawful foundation for the equal opportunities, *isonomia*, of all populations. In the ancient Athens of the fifth century B.C, urbanism had been provided with the platforms and activities that upheld democratic lifestyle with various sports and games. In ancient Greek times, an agonistic way of life was encouraged as part of political competitions in the search for the excellence (Hooper 1978, 179–92; Thomas 1993, 16). Even though the explanation of common good for forms and meanings is not easily defined (Blau et al. 1983, 300), well-designed place in the ancient Greek *polis* is not only architecturally attractive, inspirational, and pleasing, but also collectively creating local pride of their heroes, writers, statespersons, and tradition.

Urbanism and Diversity

Regarding the nature, diversity is the necessity and property of energy transformation and the web of food in the ecosystemic community. Aristotle understands *polis*, -in *Politics*, I. 1253b 20-, as a thing, which is by plan and composition to demonstrate diverse households and social classes with specific kinds of relationship. The notion of diversity, *plethos*, for Aristotle is on the necessity for individual to be self-sufficient, *autarkea* (Bresson 2015, 229). The communities or associations in the ancient Greek *polis* comprised variety of populations with specific occupations, expertise, functions, roles, fashions, forms, and characteristics. As a socioeconomic organization and political community, the Greek *polis* was by nature diverse so that its populations were able to interact, work together, and grow socially and economically for the novelty of their competitive achievements. As matter of fact, complexity and heterogeneity had been always part of the ancient Greek culture, which was famously agonistic; however, this competitiveness was directed toward the excellence and novelty; however, most the innovative achievements were advanced by the use of writing and as a result of intensified interchange of people, goods, and ideas (D'Angour 2011, 225).

The diversity of populations and their resources enabled the ancient Greeks to develop new ideas and techniques as well as knowledge and skills because they had established a system for a strong civic collectivity based on individual merit. The foundation of this civic collectivity was established by the existence of various social clubs and associations called *koinon*. In such organizations, there were not only intensive interactions and learning processes were well accommodated and facilitates but also the imposition of membership fees and donations. As matter of facts, the ancient Greeks participated and involved in various socioeconomic and cultural activities with voluntary contribution in terms of *liturgies*. Aristotle points out in *Politics* II.5 1263b 37–8 that despite the city-state is a multitude system, the populations should be united and made into a community by culture; in other words, despite each Greek *polis* was politically independent entity, they shared and upheld the coherent set of same values and ideals within the Greek language, rituals, customs, and traditions.

Aristotle points out that the cultural unity in diversity of *polis* was made possible with the Greek language through the practice of education, *paideia*. The concept of *paideia* is literally translated as educational system with apprenticeship. The word *paideia* means guide, guiding and guidance; it is from the Greek word *paidagagos (pais paidos*, boy, +*agogos*, guide).

According to Aristotle points out the importance of discourse (*logos*) and didactics (*didache*) for *paideia* that is necessary to be well-prepared for loving what is noble and hating what is base (*Politics* 1179b 29–31). *Paideia* enabled the ancient Greeks to educates, trains, and leads young citizens to overcome their plurality by means of knowledge and skills of civic culture. Nevertheless, Aristotle underscores that the nurture and exercise of young generations should be managed and regulated by law either written or unwritten (*Politics* 1180a 29 1180 b1).

Regarding city planning, Aristotle mentions that Hippodamus was the first Greek town planner who introduced straight avenues and streets in the layout of Greek towns such as in Miletus and Piraeus (Haverfield 2009, 30–3). However, the orthogonal spatial planning was evidently known in Central Italy and Western Greek cities long before the time of Hippodamus' Miletus (Winter 2006, 185). As matter of fact, the grid pattern and scheme became the spatially guiding principle for the development of urban blocks and the relationship between public and private realm. However, the grid system was not intended to divide the public and private domain but to integrate both into a sociopolitical system of *polis* through the practice of *paideia*; the educational system of the city-state worked as the parent of young generations in a systematic way. The city-state elevated and perfected the household, *oikos* (Nagle 2006, 301); it was achieved with a general educational system to establish well-mannered and highly trained citizens—*politeia*. Nevertheless, the application of the gridiron scheme was not at random but mostly for the locations with relatively flat terrain. The ancient Greek city-state of Pergamon was not designed with such gridiron system but it was provided with topographically adjusted layout by optimizing the functional defense and the stunning views to the surrounding landscapes.

Urbanism and Linkage

Spatially speaking, diversity in urbanism does not work well without good connectivity so that every part and area of the city is accessible by walk and transportation. Socioeconomically, diversity is necessary to be provided with well functioning accesses, networks, and places for interactions and communications. The existence of *polis* was established socioeconomically with a compact form and network of households and public places. Aristotle's concept of the household for polis (*Politics* 1253b 23–5) points out the importance of the skills of domestic management (*oikonomia*) that consist of functional relations and particular roles of elements within a system. The connectivity between *oikos* and *polis* is established with economically self-sufficient household of the *oikos* that enables *polis* to perform its economic viability and sociocultural civility (Nagle 2006, 31). Connectivity in the urban settlement is the hub and network system for collaborations, productions, services, recreations, and cultivations.

In regard to spatial connectivity, the Greek city-state of Miletus was one important urban layout in history. The physician Hippodamus introduced the gridiron pattern or chessboard system of streets and blocks for Miletus, Olynthus, and Piraeus in the mid of fifth century B.C. The layout with chessboard for city-states was characterized by the streets of 10–12 meters wide and blocks of 35 by 65 meters. The backbone of the streets system was the main avenue of 30 meters wide running on a North–South axis with *agora* as its urban center. On the coastal areas, the gridiron scheme provided the ancient city-states with spatial regularity, uniformity, and orderly urban fabrics and public places. The orthogonal pattern of street system became favorable design in most Greek city-states after the fourth century B.C. The introduction to geometrical layout, urban connectivity was mostly part of the expansion of the Greek culture for an efficient and effective for communication, security, utility, and mobility. Historically, the gridiron pattern provided the control over city and the transportation with oxen, horses, and mules.

Culturally speaking, the Greek language was the most important means of connectivity among the Greek communities in rural areas and city-states. Despite politically sovereign, the Greek city-states were culturally united by the Greek language, myths, and religious cults. The sense of a Greek nation was established and developed within a historical and cultural framework instead of geographical territory. The sense of cultural bond among the ancient Greeks was historically made easier with the writing system. Written system was based on the *phoenician* alphabet that had been used for records and bookkeeping since Mycenaean times; Phoenicia is a region along the eastern coast of Mediterranean Sea-. By means of the

Greek alphabet system, the Greek city-states—*polis*—became the centers of cultural, political, and economic developments with a well-established system of records and the sovereignty of law. The Draco's code was the historical law of *polis* for the justifiable homicide that made a clear distinction between premeditated and involuntarily manslaughter (Bowra 1965, 50).

Urbanism and Governmentality

The urban codes and laws provided the populations of *polis* with the sense of certitude about fairness, civic responsibility, and social justice in various activities. The religious rituals and festivals were the events for building the sense of belonging as well as the sense of civic responsibility of citizens for the ancient Greek city-states. There is no separation between secular and religious activities in the ancient Greek *polis*; participation in the rituals was highly political (Azoulay 2014, 107). In order to maintain and sustain this certitude, the *polis* was provided with governmental institutions. The first democratic government was established in the ancient Greece polis of Athens when Pericles in the fifth century B.C. (Azoulay 2014, 51–2). In guarding a good democratic governance, the *polis* was provided with the city assembly, *boule*; the member of the assembly was the representative of his community. In the assembly, every representative was an appointed person with specific task and duties such as consuls (*proxenoi*), envoys (*presbeis*), heralds (*kerykes*), and messengers (*angelou*) (Nielsen 2002, 40). Their significant role and function for the ancient *polis* were at peak when the preparations for Olympics or Pan Hellenic festivals came to take place.

As an autonomous political community, the *polis* was also an economically self-sufficient system with its own market and government. The crucial issues of the ancient city-states were peace and overpopulation. Under the leadership of Pericles, several city-states managed to sign agreement for regional and international peace and trade. Numerous *polis* worked together in a confederation or in a treaty for economic exchange and defense against non-Greek invasion and threat. Urbanity in the ancient Greek democratic Athens was architecturally recognizable by its public institutions with a religious center of Acropolis and civic center with its core at *agora*. Aristotle's in Politics points out the importance of civic buildings for urbanism, such as *agora* (market), *stoa* (public building for citizens), *theatron* (theater), and *gymnasia* (educational facility). Indeed, each city-state needed house (oikos) and the rural areas (*kome*). The ancient classical Greek *polis* was commonly provided with several monumental buildings such as *poleis* (palace or the seat of the *tyrants*), *prytaneion* (office of the oligarchic magistrates or the seat of the head of state), *bauleuterion* (city hall), *strategeton* (military headquarters), and various temples in its acropolis (Hansen and Hansen 1994, 23–89).

Even though the polis in Classical and Hellenistic periods had different government system, each *polis* was provided with a *prytaneion*, a *bauleuterion*, a *stoa*, a *gymnasia*, and an *agora*. The existence of the city-state was upheld and guarded by those institutions that maintain, cultivate, and sustain the civic socialization and prosperity of its populations. *Stoas* in the *polis* played a significant important role as an educational public building with multipurpose function. Architecturally speaking, the building of *stoa* was erected with a long open structure with a colonnade on the front side facing the marketplace. The intimate relation of public affairs between *stoa* and *agora* was provided by the ancient Greek citizens with livable civic activities because, in ancient Greek *polis*, shopping was done mostly by men (Robinson 1959, 95). Both were the centers of ancient Greek life-world. Regarding its shelter structure, *stoa* provided the citizens of *polis* with a convenient public place under shade for their casual gatherings, social interactions, and hanging out with friends (Darling 2004, 162).

In the ancient Greek *polis*, the *agora* took place on daily basis while the *bauleuterion* of *boule*—the assembly of magistrates—took place occasionally based on specific issues and certain agenda. The institution of the *boule* is a political council representing citizens and their associations—*koinonia*. Architecturally, a democratic *polis* is provided with the institutions of *bouleterion* and *stoa* that surround the marketplace, an *agora* (Dinsmoor and Anderson 1973, 263; Roth and Clark 2014, 191). The Athenian democratic assembly under Pericles took place in *bouleterion* or in *stoa four times* every month involving more than 6000 members out of 40,000 adult male populations (Williams 2009, 60). The functions of *stoa* and *agora* were essential for the sustainability and livability of any ancient Greek *polis* because of its public meetings and gatherings. One important purpose of *stoa* was to establish the codes and laws as well as to serve justice, *dikasteria*. Regarding its civic interest and participation, the execution of justice—*dikasteria*—was not always established with the courthouse. The court sessions, *dikastai*, could meet in various places; but its administration took place in the court (Hansen and Hansen 1994, 76). In doing so, the ancient democratic city-states allowed the justice sessions to be attended by large audience for openness and fairness.

Urbanism and Proximity

Density and intensity for the ancient Greek city-states were the reasonable solutions for their topographical settings. The necessity for proximity with small communities was not only geographical for defense and trade but also sociocultural for their local pride and tradition. A compact and small scale of populations were socially potential to build a collective bond among populations; unlike the Roman, the ancient Greek city-states did not have any census but the populations of Athens under Pericles were around 150,000 citizens, 35,000 metics, and 80,000 slaves (Robinson 1959, 93). Administrative, commercial, and polluting craft were concentrated in special quarters that had hygienic advantages (Bresson 2015, 45). The *polis* is an economic and political community with intensive interactions and exchanges with other city-states, regions, and colonies. The economic household of the ancient Greek does not mean a self-sufficient economic system but it should be understood as an economy based on kinship relationship with the sense of mutuality that involves empathy and not self-interest and not greed for the common good of community. Aristotle's conception of *oikonomia* does not deny the market mechanism but it asserts the ethical supremacy of the communal form (Gudeman and Hann 2015, 3).

The necessity of proximity in *polis* was to build the economic and political community that relied on and optimization its own capacity and potentiality.

Moreover, the ancient Greek *polis* promoted and valued agonistic way of life in their cultural activities. Competitiveness for human excellence and justice had been integrated into the culture of the *polis* since the classical times. In its social and spatial proximity, agonistic culture finds its way to establish professionalism and meritocracy. Moreover, proximity in the context of the *polis* is not only for the competitive endeavor but also for solidarity in the spirit of *philia* (Konstan, *Friendship in the Classical World* 1997, 67–8). In the context of the *polis*, Aristotle speaks of *philia* in the sense of a friendly bond of people in pursuit of the common good (Curren 2000, 124). Regarding its highly concentrated populations, frequent and intensive interactions were made possible to establish amenable relationship among populations; all was made possible because the supremacy of law over all people that every person had a civic duty and responsibility to do their best. Hence, the Aristotelian *philia* is a kind of spiritual brotherhood of humankind that built a civic collectivity with joyful engagement.

One important occasion of *philia* was exercised in the activity of *symposion*, which means wine drinking party among adult male people. The culture of *symposion* was well demonstrated and recorded by various ancient Greek pots (kraters) or mixing bowls and cups. Most illustrations of *symposion* feast were associated with the Greek mythical heroes and their everyday life. The *symposion* took place after the participants chose their chief of conversation. The ancient Greek *symposion* was the platform of the citizens of *polis* to discuss serious matters with socially relaxed atmosphere; this was also the informal forum of education and sharing of experiences. In most parts of *symposion* was the recollection of daily life. The proximate demography of *polis* enabled them to know each other better through festivals, rituals, and other civic events. However, all their collective activities were not only liturgies but also entertainments. The *Odeon* served the populations with musical performances, while the *theatron* was the place to behold the life-world through the Greek myths, comedic, tragic, and satiric dramas. In the ancient Greek dramatic plays, from Antigone of Sophocles to Illiad of Homer, the apprehension of death and anguish were performed over and over again to remind people of solidarity, civil courage, and empathy that bring their collective awareness for their *polis*.

Urbanism and Local Identity

Local identity in the context of ancient Greek *polis* was indispensable and important for the sense of home for the Greeks. The establishment of this identity was historically constructed through experiences of wars and liturgies as well as festive competitions among city-states. Local identity for the ancient Greeks was commemorated with the statues of their heroes at the *agora* and other civic places. All citizens of the ancient Greek *polis* have the civic duty and rights of liberty according to laws and decrees; they commemorated their own obligations and performances in the epigraphy and iconography of their burial markers (Liddel 2007, 310). The things they shared for local identity were architecturally and culturally immortalized in their collective domain (*chora*) and sociocultural association (*koinon*); each member of koinon was proud of their home place and group. The proof of their membership in such a political community is paying tax (Orrieux and Schmitt 1999, 329–31).

For Aristotle (1332a 25–7), there are nothing good things lacking the good environment, *choregia*. Henceforth, *polis* is the conducive environment for the ancient Greeks to be *politeia* because of a well-administered system (*eunomesthai*) and architecturally well-designed (*choregia*). The unity of civic culture and the built environment is necessary for urban policy for the sense of locality. In other words, the greatness of the ancient Greek *polis* represented its citizens and their sociocultural and political achievements. Doric, Ionic, and Corinthian architecture become parts of the Classical Greek dialects and traditions in the late of seventh century B.C. The epics of Homer provided the ancient Greeks with collective cultural bond.

Culturally speaking, each *polis* has its own customs and traditions. Most of them were representative by public buildings. Improvement, refinement, and enhancement at the pediments and entablatures are common practice at every *polis*. The best example is the Parthenon of Athens (447–432 B.C.), which was constructed under Pericles' political leadership. Ictinus and Callicrates were the master builders for this temple while Phidias worked as the leading sculptor. Architecturally, the numbers and distance of architrave vary from one temple to another one. The refinement of the building with well-proportioned composition was considered delicate and perfect from various distances of points of view. As matter of fact, the Doric and Ionic architecture have achieved its ultimate enhancement for overcoming optical illusions and perfect proportion of Athenian public buildings in the end of the fifth century B.C.

Besides Architecture, the ancient Greek *polis* populations were passionate appreciators of Greek dramatic tragedies of Aeschylus, Sophocles, and Euripides as well as the enthusiastic participants of Pan Hellenistic sports competitions. The public buildings such as *theatron* and *stadion* were always highly occupied by the crows with open-air performances during all seasons. Until today, the architecture of *theatron* is remarkable buildings with its integration into the dramatic topographical setting of the landscape, such as the *theatron* of Epidaurus, which was built in Delphi in circa 350 B.C. This building is an extraordinary example of ancient Greek ingenuity for performing arts with an extraordinary audiovisual quality. The *theatron* is architecturally supported with a stage called *orchestra* (meaning literally dancing floor), and a backstage building called *skene*. As a whole, the ancient Greek *theatron* was the platform of *polis* in sustaining ancient Greek agnostic virtues of nobility and civil courage.

Stadion was one of the other public buildings that upheld and sustained the ancient Greek culture. The athletic contests and festivals took place in every *polis* that upheld the agonistic values system. Unlike *theatron* of performing arts, *stadion* was the stage and spectacle of actual performances of the fierce contest on strength, courage, endurance, and excellence. The Olympic festivals and Pan Hellenic sports were the events when all ancient Greek populations gathered and shared their joys and ideals, beyond the *polis* boundary. In addition to the periodical festivals, there were some civic feasts that celebrated by the ancient Greek populations regarding the visit of king and the attendance of assembly's members (Strootman 2018).

Concluding Notes

The question of urbanism is inevitable to count on local culture and its history. Regarding the importance of collective achievements in knowledge, beliefs, technology, arts, and literature; local knowledge is always contextual from geographical settings and its acculturation based on internal and external exchanges of ideas, traditions, and customs. Since culture is a text written by local people (Geertz 1983, 50), original concepts in its own language are the essential sources of experience, wisdom, and knowledge. Despite historical discrepancies, the ancient Greeks in the fifth B.C. Athens and in the modern democratic societies have something in common concerning the intentionality of urbanism, namely civility. Urbanity implies the lifestyle that regards the livability of public realm beyond automobile. However, one obvious challenge of contemporary urbanism is to establish home with pedestrian-oriented environment (Beatley 2005, 278).

Urbanism with democratic governmentality is considered one of the best options of civilization. This option is good because it upholds the individual freedom based on the supremacy of law and order for social justice and equal opportunity for all. However, the foundation of democratic urbanism demands a democratic economic production as well. In doing so, the resources, especially the capital of productions and services, are not exclusively dominating. Democratic urbanism implies political and economic governmentality with the balance of rights and responsibilities of all individuals in the system of productions and services. The democratic system of economy is considered not yet achieved neither by the ancient Greek Athens in the fifth century B.C. nor by the modern societies. Until today, the economy of urban centers is predominantly controlled by the globally operating big corporations and financial institutions. The task of urbanism today is not only to cultivate our political behaviors and transform our lifestyle from automobile dependent to pedestrian-oriented routine. Democratic system of economic productions and services is considered necessary for urbanity because it eradicates inequality of populations based on their wealth; under the current market capitalist economy, the rich gets richer while the poor becomes poorer because the return of investment of the capitals is always greater the growth of economic production.

References

Aristotle, and Richard Kraut. 1997. *Politics.* Oxford: Oxford University Press.
Azoulay, Vincent. 2014. *Pericles of Athens.* New York: Princeton University Press.
Barker, Ernest. 1948. *Traditions of Civility: Eight Essays.* Oxford: Oxford University Press.

Barker, Sir Ernest. 1958/1998. *The Politics of Aristotle*. Oxford: Oxford University Press.

Beatley, Timothy. 2005. *Native to Nowhere: Sustaining Home and Community in a Global Age*. Washington, DC: Island Press.

Blau, Judith R., Mark La Gory, and John Pipkin. 1983. *Professionals and Urban Form*. Buffalo: SUNY Press.

Bowra, C. M. 1965. *Classical Greece, Great Ages of Man*. New York: Time Life Books.

Bresson, Alain. 2015. *The Making of the Ancient Greek Economy: Institutions, Markets, and Growth in the City-States*. New York: Princeton University Press.

Cowan, Alexander, and Jill Steward. 2007. *The Cities and the Senses: Urban Culture Since 1500*. Surrey: Ashgate.

Curren, Randall. 2000. *Aristotle on the Necessity of Public Education*. Lanham: Rowman & Littlefield.

D'Angour, Armand. 2011. *The Greeks and the New: Novelty in Ancient Greek Imagination and Experience*. Cambridge: Cambridge University Press.

Darling, Janina. 2004. *Architecture of Ancient Greece: An Account of Its Historic Development*. New York: Biblo & Tannen.

Dinsmoor, William Bell, and William James Anderson. 1973. *The Architecture of Ancient Greece: An Account of Its Historic Development*. Cheshire, CT: Biblo & Tannen Publishers.

Doxiadis, Constantinos A. 1964. "The Ancient Greek City and the City of the Present." *Ekistics* 18 (108): 346–64.

Eriksen, Trond Berg. 1876/1976. *Bios Theoretikos: Notes on Aristotle's Ethica Nicomachea X, 6–8*. Aarhus, DK: Universitetsfod.

Fitzgerald, Joseph. 1901. *Word and Phrase: True and False Use in English*. Chicago: A.C. McClurg.

Fitzgerald, Timothy. 2007. *Discourse on Civility and Barbarity: A Critical History of Religion and Related Categories*. Oxford: Oxford University Press.

Garver, Eugene. 1994. *Aristotle's Rhetoric: An Art of Character*. Chicago: University of Chicago Press.

Geertz, Clifford. 1983. *Local Knowledge*. New York: Basic Book.

Gernet, Louis. 1981. *The Anthropology of Ancient Greece*. Baltimore: John Hopkins University Press.

Gonzales, Francisco. 2006. "Beyond of Beneath Good and Evil? Heidegger's Purification of Aristotle's Ethics." In *Heidegger and the Greeks: Interpretative Essays*, edited by Drew Hyland and Panteleimon Manossakis, 127–56. Bloomington: Indiana University Press.

Gudeman, Stephen, and Chris Hann. 2015. *Oikos and Market: Explorations in Self-Sufficiency After Socialism.* New York and Oxford: Berghahn.

Hakim, Besim S. 2014. *Mediterranean Urbanism: Historic Urban/Building Rules and Processes.* Dordrecht: Springer.

Hansen, Modens Herman. 1993. *The Ancient Greek City-State: Symposium on the Occasion of the 250th Anniversary of the Royal Danish Academy of Sciences and Letters July, 1–4 1992.* Copenhagen: Kongelige Danske Videnskabenes Selskab.

Hansen, Mogens Herman, and Tobias Fischer Hansen. 1994. "Monumental Political Architecture in Archaic and Classical Greek Poleis." In *From Political Architecture to Stephanus Byzantius,* edited by David Whitehead, 23–89. Stuttgart: Franz Steiner.

Harris, Nathaniel. 2004. *History of Ancient Greece.* New York: Barnes & Noble.

Haverfield, Francis-John. 2009. *Ancient Town Planning.* Auckland, New Zealand: BiblioLife LLC.

Haworth, Alan. 2004. *Understanding the Political Philosophers: From Ancient to Modern Times.* London: Routledge.

Heidegger, Martin. 1991. *Nietzsche Volume I & II.* Edited by David Farrell Krell. Translated by David Farrell Krell. San Francisco: HarperSanFrancisco.

Hooper, Finley. 1978. *Greek Realities, Life and Thought in Ancient Greece.* Detroit: Wayne State University Press.

Jaeger, Werner Wilhelm. 1939/1986. *Paideia: The Ideals of Greek Culture.* Translated by Gilbert Highet. Oxford: Oxford University Press.

Jencks, Chris. 2004. *Urban Culture: Critical Concepts in Literary and Cultural Studies.* London: Routledge.

Jordan, Michael. 1993. *Ancient Concepts of Philosophy.* London: Routledge.

Julier, Guy. 2000. *The Culture of Design, Culture, Media and Identities Series.* London: Sage.

Kalimitziz, Kostas. 2000. *Aristotle on Political Enmity and Disease: An Inquiry into Stasis.* Albany NY: SUNY Press.

Konstan, David. 1997. *Friendship in the Classical World.* Cambridge: Cambridge University Press.

Krell, David Farell. 1988. "Knowledge Is Remembrance; Diotima's Instruction." In *Post-structuralist Classics,* edited by Andrew E. Benjamin, 160–72. London: Routledge.

LeGates, Richard T., and Frederic Stout. 2003. *The City Reader: Routledge Urban Reader Series.* London: Routledge.

Liddel, Peter. 2007. *Civic Obligation and Individual Liberty in Ancient Athens.* Oxford: Oxford University Press.

Lynch, Kevin. 1981. *A Theory of Good City Form*. Cambridge: The MIT Press.

Martin-McAuliffe, Samantha L., and Daniel M. Millette. 2017. *Ancient Urban Planning in the Mediterranean: New Research Directions*. New York and London: Routledge.

Moller, Astrid. 2000. *Trade in Archaic Greece*. Oxford: Oxford University Press.

Morris, Ian, and Barry B. Powell. 2010. *The Greeks: History, Culture, and Society*. 2nd ed. Englewood Cliffs: Prentice Hall.

Nagel, Mechthild. 2002. *Masking the Abject: A Genealogy of Play*. Lanham: Lexington.

Nagle, D. Brendan. 2006. *The Household as the Foundation of Aristotle's Polis*. Cambridge: Cambridge University Press.

Nielsen, Thomas Heine. 2002. *Even More Studies in Ancient Greek Polis*. Stuttgart: Frans Steiner Verlag.

Ober, Josiah. 2011. *Political Dissent in Democratic Athens: Intellectual Critics of Popular Rule*. New York: Princeton University Press.

Orr, David. 2004. *The Nature of Design: Ecology, Culture, and Human Intention*. Oxford and London: Oxford University Press.

Orrieux, Claude, and Pauline Schmitt. 1999. *A History of Ancient Greece*. London: Blackwell.

Pagden, Anthony. 2002. *The Idea of Europe; from Antiquity to European Union*. Cambridge: Cambridge University Press.

Preston, Laura, and Sara Owen. 2009. *Inside the City in the Greek World*. Oxford: Oxbow.

Price, Jonathan J. 2001. *Thucydides and Internal War*. Cambridge: Cambridge University Press.

Robinson, Charles Alexander. 1959. *Athens in the Age of Pericles*. Norman: University of Oklahoma Press.

Roth, Leland M., and Amanda C. Roth Clark. 2014. *Understanding Architecture: Its Elements, History, and Meaning*. Boulder: Westview.

Schmitt-Pantel, Pauline. 1991. "Collective Activities and the Political in the Greek City." In *The Greek City: From Homer to Alexander*, edited by Oswyn Murray, 199–213. Oxford: Clarendon Press.

Strootman, Rolf. 2018. "Feasting and Polis Institutions." In *Leiden*, edited by Floris van den Eijnde, Josine Blok, and Rolf Strootman, 273–86. Leiden and Boston: Brill.

Taylor, Claire. 2015. "Social Networks and Social Mobility in Fourth-Century Athens." In *Communities and Networks in the Ancient Greek World*, edited by Claire Taylor and Kostas Vlassopoulos, 35–53. Oxford: Oxford University Press.

Taylor, Claire, and Kostas Vlassopoulos. 2015. *Communities and Networks in the Ancient Greek World.* Oxford: Oxford University Press.

Thomas, Carol. 1993. *Understanding Architecture: Its Elements, History, and Meaning.* Oxford: Westview Press.

Voegelin, Eric, and Dante L. Germino. 2000. *Order and History, Volume III: Plato and Aristotle.* Columbia: University f Missouri Press.

Williams, Jean Kinney. 2009. *Empire of Ancient Greece.* New York: Facts on File.

Winter, Frederick E. 2006. *Studies in Hellenistic Architecture.* Toronto: University of Toronto Press.

4

Urban Planning and Development

The Origin and Development

Planning has been well known as a systematic way and tool to achieve a goal. In the context of urban and regional development, planning plays an important role in identifying, allocating, determining, preserving, and conserving the use of land and its resources. The purpose of planning is twofold: transforming and stewarding the environment so that enable people to utilize all resources—land, water, flora, and fauna—within without the depletion of its capacity and quality in the future. Historically speaking, the birth of modern urban planning has been inseparable from the emergence of industrial society in various places, especially in England, Europe, and North America. Studies, theories, and practices on the evolution of planning as a system of knowledge and practice have been well documented and developed by numerous individuals such as Ebenezer Howard (1850–1928), Raymond Unwin (1863–1940), Barry Parker (1867–1947), Clarence Perry (1872–1944), Patrick Geddes (1854–1932), Patrick Abercrombie (1879–1957), Frank Lloyd Wright (1869–1959), Le Corbusier (1887–1965), Christopher Alexander (1979, 1987), Leon Krier (Houses, Palaces, Cities 1984), Peter Hall (1975,

© The Author(s) 2020
B. Wiryomartono, *Livability and Sustainability of Urbanism*,
https://doi.org/10.1007/978-981-13-8972-6_4

2001, 2014), Peter Calthorpe (1993), and Andres Duany et al. (2000, 2010).

History of planning cannot overlook its essential relation to human geography. In other words, planning is not impartial from its land and people. Every location has its own planning system and process because every location has its own specific geography in terms of issues, constraints, and opportunities. Planning becomes urgent and necessary in response to rapid population growth, the inadequate area of land, and resources. The main problem of spatial and physical planning is to build a strong, safe, healthy, and sustainable community, from the neighborhood, quarter/ward, township, city, to metropolitan, and region. In this sense, planning is understood as a process and system to integrate people and place into a world we call it a home in a sustainable way; the integration includes economic, environmental, and sociocultural livability and sustainability. In other words, planning in the context of the built environment is to build communities with specific technical standards of public safety, health, and environmental sustainability.

Community Planning and Urbanism

The planning system for urbanism is not in a way of making a blueprint of future development. As a matter of fact, such a blueprint never works for several reasons because no one is able to predict and forecast the future with a grand design. What people do for urban development is a learning process through trials and errors. However, this process is by no means an experiment but an attempt with a clear vision concerning urbanity. Urbanism has a vision that is about the culture of working, living, and playing together in a highly concentrated settlement. Planning for urbanism is to build and develop a culture, which is contextual. The planning system is contextual in terms of geography, culture, and history; in this chapter, urbanism in Canada is chosen as an example. However, it is necessary to put the planning system into a theoretical framework for urbanism. In this point, this chapter will set forth theoretical possibilities that enable plans to work for urbanism.

Studies on the Canadian planning system and practices have been presented by a number of scholars (Hodge and Gordon 2008; Thomas 2016;

Cullingworth 1987; Grant 2006). However, this study is an attempt to fill out the gap between theory and practice of planning with a focus on the plans for urbanism. Theoretically speaking, urbanism in Canada is never out of the commission of community planning. The concept of community is understood as an imaginatively constructed bond of populations that work, live, and play together because they share something in common for their daily needs and their future opportunity. Urban community is distinguished from village one not only because of the complexity and number of populations but also that of the existence of anonymous crowds in public places. By the definition of Statistics Canada, urban community is a human settlement with a concentration of population more than 1000 people and density minimum 400 populations in one square kilometer.

The community plan in Canadian context plays an important role in governing a community from an urban area, small town to city, metropolitan, and region. Community planning is likely the most appropriate term to use in Canadian context regarding its urban geographical setting (Hodge and Gordon 2008, 13); Canadian urban settlements are mostly dominated by small settlements and its sociocultural values system of collective activities and participation. Under the notion of community planning, the physical development of the land is practically associated with inhabitable property and its supporting areas for human settlement. However, plans do not work without viable goals for urbanism. In the broadest sense of the word, the community planning for urbanism is a systematic plan of settlement that works to build a home of people with a high density of diverse populations. Such a settlement should be integrated into a network of public transportation and utilities. In terms of urbanism, the urban settlement is not necessarily established and developed with a community planning. However, community planning is considered a systematic way and tool to improve and enhance the quality of life for an urban settlement.

The key component of the planning system is the clear status of the land in terms of ownership and its land use designation. In Canadian context, the status of the land is essential that belongs to personal or corporate property; the law of the country has established the relationship between a legitimate community interest and the use of land. In this line of thinking, the value of the land is not an absolute private commodity. Planning is one important instrument to manage and conserve the use the land and the layout of buildings wisely so that in any development of the land there is a legitimate community interest in their safe, healthy, and attractive environment.

Nevertheless, the community planning for urbanism is to ensure that the use of the land is necessarily appropriate for high-density habitation and supported by a good system of transportations, utilities, and infrastructure. Since to plan implies deliberately to impose humanly ordering system into the natural environment, planning practice should be held accountable for preserving and conserving the land and its resources. Moreover, the legitimate community interest includes the spaces for physical and sociocultural infrastructure and nature conservation as well as for natural disaster prevention and mitigation.

Land

Historically speaking, the site of urbanism is not restricted by its geographical setting and characteristics; this can be in the coastal areas or in the mountainous hinterlands. The most important thing for the urban site is the availability of water resources. Geologically, the water supply consists of surface and underground resources. Since ancient times, the problem of water resources for urban habitation is not only about to find the water wells or water springs but also to manage the use and distribution of this essential resource. Urban planning is necessarily aware of this problem. The community planning is necessary to be in alignment with the regional and environmental planning in the region. This alignment is to secure the upstream areas, watersheds, open green spaces, and forests into a natural conservation region that upholds and protects the surface and groundwater supplies to the adjacent urban settlement. In other words, the relationship between the buildable and unbuildable area is necessary to be established as a whole ecological system in a proportional manner so that the water conservation is possible and sustainable.

Besides the water supplies, the land and its legal status are indispensable for community planning. Regarding this essential nature, Canada launched a land inventory program between 1965 and 1975. The program was able to assess and record covering over 2.5 million square kilometers of land and water. Accordingly, over 1000 map-sheets at the 1: 250,000 scale were published until the early 1980s. Even though updating and renewing the information is necessary so that better information is available for several areas as part of more recent soil surveys, most data of the maps are still mostly effective, and many areal jurisdictions still useful for the purpose of land use planning. The aim of this inventory program is to identify the ecologically bearing capacity of the land and its resources for agricultural and other feasible activities.

The land in Canada is owned by individuals on a statutory basis, but not on the constitutional right. The ownership of land does not include in the Canadian Charter of Rights and Freedoms. The government is able to acquire the land without the owner's consent with the Expropriation Act; in the United States, this is called as the 'condemnation' under the right of 'eminent domain', in the UK known as the 'compulsory purchase'. However, the government should follow the law as to what land may be expropriated and must perceive the measures set out in the legislation that commonly serve to protect the private landowner. The federal or provincial government can expropriate the land without the consent of the owner. However, most appropriations should go through the provincial applicable legislation. Based on the land inventory, most areas are under the administration of the Crown; the Crown land is the term for the land owned by the federal or provincial governments. Only less than 12% of the land in Canada belong to private owners. In certain regions, such as Yukon, Nunavut, and the Northern Territories, the land is administered on behalf of Canada by the Aboriginal Affairs and Northern Development.

The importance of map is to identify and document the land and its geographical features, habitats, and resources. Planning needs the map of the land that enables us to see the possibilities and constraints of the location. Today, a geographic information system or GIS is widely applied for spatial and land use planning for analysis, modeling, and visualization. This is a system designed to capture, store, work, analyze, manage, and present various geographical data in digital form, platform, and application. The GIS provides the information of a location with the Cartesian coordinate on the global system. In doing so, every location and movement on the globe can be recorded and monitored with digitally mapping system in raster or vector image. By handing out geospatial data from the satellite imaging, remote sensors, and aerial photography planners obtain a detailed perspective on land, utilities, and environment. Technologically speaking, the GIS enables planners from various disciplines: land use, transportation, environment, land surveyor, geography, geology, and engineering to work together on the same platform and database. The use of the GIS becomes essential and referential for various professions and businesses that demand the geographical data and maps. However, field survey is always necessary to collect and check the data for more accurate findings. The topography of the terrain is mostly in need of field survey for more detailed information. There are the common applications for GIS that comprise: evaluation and analysis of plans for urban development, examination on the regulatory compliance, preservation of historic areas, environmental impact assessment, regional planning surrounding the city, and mapping the areas for special occasions, temporary closures of utilities, and other emergency purposes.

Planning Act or Law

Historically speaking, the birth of urban planning in England, Germany, and Canada began with community movement for the conservation of nature. In this tradition, planners are reminded to be aware of the areas where they are ecologically not able to build for human settlements. The establishment of the planning system in this country was motivated and generated by the community's movement, which was inspired by the Garden City Movement in Great Britain and City Beautiful in the United States. The outcome of community's movement for nature conservation and the healthy neighborhood was the establishment of planning boards or councils in various towns and cities in four provinces: Ontario, Nova Scotia, New Brunswick, and Alberta; these four provinces had been able to establish the planning statutes. By 1925, every province, except Quebec, was provided with the similar planning act. The purpose of the act is to provide the municipalities and regions with the statutory power to regulate and manage the physical development in their area. However, the planning act is not mandatorily intended to put the municipalities or regions in active role and function for providing physical plans. The intention of the planning act or planning law is to enable the provincial or local government to manage and direct the development with a policy statement. Generally speaking, the planning act or law is the legal framework to manage the physical urban development with regard to the principles of compact, safe, secure, healthy, and attractive urbanism.

In Canada, the planning acts of various provinces have been evolving and developing with periodical amendments at the end of the World War II. All these are to update and engage with current issues concerning community, environment, and economy. One important regulatory system of urban development is zoning by law. This regulatory system was adopted and adjusted by the municipalities in Canada from the American experience in the cities of Los Angeles in 1908 and New York in 1916. Kitchener was the first Canadian municipality implementing the zoning ordinance in 1918. The main idea of a zoning ordinance is to regulate the use of land with a certain function, population density, and building height. The objective of this regulatory system is to control the physical development in the urban area for the safe and healthy environment. Restriction of building height and setback were enacted in some big cities that were in order to allow sunlight to reach every building in a neighboring complex. The zoning was also an effective regulation in preventing industrial activities in encroaching residential and commercial areas. In Canada, the practice of zoning is regulated by the municipalities and regions under the provincial legislation, development policy, and plans.

Moreover, the planning act or law is to establish a room of interaction and relationship between private and public interests and initiatives in the land development during the planning and design process. The planning act provides the municipalities and regions with the legislation of the system, process, and procedure of how development policy is implemented to frame the needs and interests of development within the policies of land use development and the regulatory applications of the permit. The policies of development encompass legislations, development guidelines, and strategic plan at the provincial level, as well as official plans, zoning by law, planning decisions, and building permits at the municipal and regional level of government. The applications of development include the site plan application, subdivision/condominium development application, and building permit.

As with any other discipline, planning has been historically evolving from a simple need to a complex one. In Canada, the Planning Act has been playing an important role in providing the development of the land since the beginning of the twentieth century in various provinces; the act establishes the existence of planning as discipline and profession. The Act permits municipalities in various provinces to prepare their city planning visions, schemes, and regulations concerning the built environment. The first British Planning Act of 1909 was a model for planning framework for healthy and attractive communities in Canada. In each municipality of Ontario, there is a planning board of council that works to mediate and facilitates local aspirations and concerns in response to the private developers' plans. The Ontario Municipal Board was such an administrative board consisting of independent members. The function of the board was to hear and to resolve the disputes of planning and development issues between private developers and the municipal or regional government. The board reported to the Ministry of Attorney General. All members of the board are appointed by the provincial government. The members are professional planners, lawyers, former city officials, and professionals with various backgrounds, such as economics and business. The appointment of members was commonly for terms of two to five years with a maximum of 10 years in a position. Since its establishment in 1906, the board had mostly operated as a tribunal of disputes between municipalities and private developers concerning development issues, such as historic conservation, architectural scale, height, and volume, and community's related values. Over last decades of operation, the board had been heavily criticized by citizens and community's interest groups that this planning body was weak in representing local interests and concerns. From the public opinion, the provincial planning board was considered less democratic and inefficient in its operation (Pagliaro 2017). Since April 3, 2018, the provincial planning board was functionally replaced by local planning board with less scope and limited power.

As a legal framework system of planning and development, planning act or law enables the public and private sector to establish, actualize, and develop urbanism. In order to achieve the goals of urbanism, the act or law should be accommodative and flexible to deal with the current and future issues of economy, environment, and society. However, the focus of the statements in the act or law should be solid and clear for engendering the safe, healthy, and enjoyable urban life-world. The statements guide and direct the decisions and actions of development within the framework of urban livability and sustainability in terms of economy, ecology, and sociology. In its mission, the statements of the act or law should be viable and measurable for practical actualization in dealing with the local and global conditions. Thus, the planning act or law includes a system of strategic and statuary of elements, procedures, resources, relations, and activities of planning and development for urbanism.

Plans

Plans are the forms, documents, and deliverables of planning that works to direct, manage, and control of urban development according to certain principles. Urbanism provides the principles for the safe, healthy, and attractive built environment with certain requirements of the density of populations, land use destinations, and diverse scales, forms, and types of places, streets, pedestrians, residential, commercials, industrials, and workplaces. In most cases, plans make ideas, requirements, criteria, and expectations of urbanism to be tangible and measurable; in doing so, we are able to estimate how many resources are needed and utilized for the actualization and adjustment of the plans. There are at least three essential categories of plans for urbanism. The first category consists of strategic policies, guidelines, and regulations for the whole urban area. The strategic plans encompass essential aspects and components of urban development of the whole area of town or city in dealing with a long-term validity, compatibility, and outlook.

The strategic plans are potential to characterize the urban built environment and its public realm activities with certain features of architectural skyline, composition, form, style, performance, and detail. In terms of urbanism, the strategic plans are indispensable to integrate the components of urbanism, such as high-density habitation, mixed uses, public transit system, energy conservation, regional and global connectivity, into a compact settlement that is conducive for economic growth and safe, healthy, and attractive man-made environment. Substantially speaking, the strategic and long-range plans should contain the principles of viable schemes and workable programs for the livability and sustainability of urbanism. One important aspect of these principles includes the general policy to establish and develop the safe, healthy, and attractive walkable public realms in every district and neighborhood in the city.

The second category of plans concerns the specific policies and regulations in certain districts or subdivisions; these plans include requirements, regulations, and guidelines of development in terms of rehabilitation, renovation, restoration, and revitalization in a parcel or more sites. The second category includes the treatment of the architecturally unifying elements and structure for local character and identity. Most importantly, the second category of plans is to engender economic and social livability of a district. However, such a livability is necessary for a well-connected integration with other districts by public transit system and other public utilities. The mission and goal of the second category of plans are to establish and develop subdivisions or districts with distinct activities of productions, services, and recreations that uphold diverse functions of the whole urban settlement in the web of interdependent economic, ecological, and sociocultural system.

The third category of plans covers the plans for a location in a district or an urban area; this kind of plans is to ensure and establish the relationship between the development on the site and the surrounding in dealing with urbanism. In its practical sense, the third category of plans shows the guidelines of the potential interaction between the inside and the public realm. Moreover, the plans should include the treatments of building bulk, entry, emergency access, parking, fire route, and landscaping. In the context of a whole system of urban settlement, the mission and goal of the third plans are to establish and develop the location and building system for strengthening district identity and uniqueness with specific activities and architectural endeavor.

From its beginning, physical planning of land use in Canada had been the business of municipalities. The planning act provides the municipalities with the power to prepare a general plan or the official plan. This plan contains general and comprehensive directions and principles of physical development and land use. The purpose of the municipality's official plan is: to regulate the use of land, to evaluate and settle the potentially conflicting land uses of local, regional, and provincial interests, to coordinate the use of land for infrastructure and public utilities, to manage, to provide the municipality with the framework of zoning by law for the density of habitable land and building height, and to highlight the future growth of the areas. The official plan is statutory for a five-year period; the municipal council should review, amendment, and update the plan every five years. Prior to its coming into effect, the official plan should be ensured that has held at least one public meeting with a public notice in various media at least 20 days ahead of the time, is consistent with the provincial policy statement, must be consulted and approved by municipal council, local planning board, and provincial planning authority. The review of the official plan is held in a public meeting with an open house, in which any citizen or public body may write their comment or speak at the public meeting about the anticipated plan.

The zoning bylaw is to put the official plan into effect and to provide the management of land use for its daily business. All new development applications should comply with the zoning for a building permit. Zoning by law is a set of regulations and requirements for the use of land and the layout of parcels, buildings, and structures. The purpose of zoning by law is to specify the land use with what is and not allowed to be built on the land. In doing so, the zoning bylaw prevents potentially conflicting and incompatible land uses in a community. However, there is an opportunity for a zoning transformation for a new development project. This process may be applied as rezoning or zoning bylaw application as long as it is in the alignment of with the official plan. The amendment of the zoning is subject to the approval of the municipal council. In case, the council turns down the application, the application may appeal to the planning board; this body is an independent administrative tribunal that is publicly accountable for hearing appeals and making decisions on contentious municipal affairs. Prior to its enactment, the zoning bylaw should hold at least one public meeting and open house. The zoning by law in Ontario is commonly required to be reviewed and updated not less than three years after the approval of the official plan five-year review. How flexible is the zoning by law? The planning act grants the municipality to appoint a committee of adjustment for minor variance of the zoning by law. The members of this committee are elected from citizens. The committee has the power to decide on minor modifications from the provision of the zoning by law as well as the permission for the extension and enlargement of building a structure that is legally non-conforming or an alteration in a non-conforming use.

The site plan permit is required for a specific area of new development prior to the issuance of building permit for all developments subject to the site plan control. Such an area is chosen by the municipality with special planning consideration because of specific reasons in relation to the provincial policy statements and the municipal official plan. Site plan review and approval are mandatory for the development proposals occurring in the mature areas that includes the additions of the new structure. Site plan control is a planning tool for managing the physical development in a certain area for the architectural character, aesthetics, economic growth, the livability of the public realm, and environmental conservation. However, some municipalities in Ontario set their own preferences for the areas under the site plan control review and approval. A site plan by law enables the municipality to approve the design and other technical aspects of a proposed development. The purpose of the site plan by law is to integrate a new development into a whole architectural and environmental system of urban block or neighborhood. This integration is commonly related to the improvement and enhancement of architectural and environmental quality of the pedestrian network, streetscape, public places, and green spaces.

A draft plan approval is required for a new development when a site is divided into two or more parcels that may contain a significant number of roads, open green areas, and public utilities. The municipal approval includes for site plan permit and subdivision plan authorization. The requirements of the subdivision permit in the City of Toronto include the inventory documentation, studies, models, the conceptual site plan and landscaping, the construction management plan, the environmental assessment of the site, and the geospatial context of location. The inventory documents encompass the layouts of boundaries, easements, reserves, circulation, general and accessible parking, driveways, above and below grade public utilities, landscaping, grading and retaining walls, and lighting. Meanwhile, the studies comprise community and facility studies that identify the current and required social infrastructure to uphold the safety and health of prospective inhabitants. For the more complex site of five hectares or larger site, there is a long list of studies for subdivision approval such as studies of noise impact, flood and erosion control, hydrogeological and geotechnical condition, pedestrian wind level, sun shadow, topography, and traffic operations assessment. Such studies are to ensure that the proposed development is an integrated part of development policy for socioeconomic livability and environmental sustainability. In order to achieve the integration of a new development into a whole vision of the official plan, municipalities commonly provide development guidelines in terms of architectural and constructional principles, schedules, and documentation.

A new development in an urban area requires a building permit. The permit includes a set of conditions and requirements for safe and healthy construction. All these are mostly described in the municipal building code. In Ontario, a building permit is required for a new construction on the land with more than ten square meters. This requirement includes renovation, repairs, an extension to an existing structure, the excavation of new foundation, and the construction of a seasonal building. The purpose of the building code is to set the standards of safety, health, fire protection, flood prevention, accessibility, energy efficient, environmental sustainability, and resource conservation. All requirements of the building code are necessarily in compliance with the local zoning by law and other planning regulations, such as site plan and building control. Site plan, floor plans, and sections of the building are essential drawings for building permit application.

Nevertheless, plans are not workable visions, without definitive goals, and viable policies. Vision in planning is essential that enables us to see the quality of the world to be sought. Vision enables us to bring about the planning goal tangible so that we are able to see the quality of the urban life-world we dream on and seek. In this sense, the world is not understood simply about the totality of things in the built environment. Rather, the world is conceived as the beings as a whole. This includes the people who live, work, and play in the certain area that builds a strong sociocultural and economic bond for their livability and sustainability in the location; this is the life-world where the home is. Accordingly, the plans for the life-world are not simply the schemes, projects, principles, and documents to manage the development. Rather, the plans are to care and work for the life-world. The plans are to engender, organize, and facilitate the growth of environmental, economic and sociocultural quality of the community's world. In the broadest sense of the word, plan in the context of urbanism is not only the tool of development but it is a system of visions, objectives, and policies to establish, develop, and maintain the community's world based on their collectively shared values system. Engagement and participation of all members of the urban community are essential and important so that urban development is by all, with all, and for all populations.

A strong collective vision is necessary for such development that enables people to know the direction and outcome of all plans. In order to grasp the vision with tangible and measurable things, the vision of plans needs viable objectives concerning various aspects, such as building performances, architecture, landscape, public facilities and infrastructure, transportation and transit system, energy, sewage, and waste management, economy and business, culture and tourism, etc. From these objectives, there are the preferred courses of action to be followed; these courses are called policies. Urban development policies seek to address issues, constraints, and opportunities regarding populations, energy, environment, industry, housing, transportation, employment, health, public safety, social condition, etc.

Prior to set up plans, it is necessary to collect data of demography, geographic location, history, and all sociocultural, economic, and environmental resources of the site. The demographic data stand out for community planning because the composition of diverse people with their potentialities and activities determines specific issues, constraints, and opportunities for their current and future world. However, identifying, compiling, reading, and studying the demography of a location and its socioeconomic and environmental setting is preparatory work of community planning. The purpose of this preliminary work is to seek the possibilities of various activities into a system of productions and services. This is to integrate work, live, and play into the land use scheme and network. The integration of work, live, and play activities in the urban life-world entail a comprehensive understanding that urban settlement is a home. However, in order to hold the sociocultural and economic livability of the location, the urban built environment is necessarily characterized by a significant portion of the residential function. In other words, the urban life-world entails a dominant area of residential that potentially upholds a safe, healthy, and attractive place to stay and grow together as an economic and sociocultural community.

Housing is one of the essential aspects of urban settlement that upholds the stability, livability, and sustainability of the socioeconomically urbanized world. In Canada, Canada Mortgage and Housing Corporation (CMHC) have been playing an important role in shaping and establishing the quality of housing in the country since 1946; this is the state-owned corporation of Canada with specific mandate for assisting all Canadians with affordable housing mortgage and providing the government and industry with data of research on housing on periodical basis. Moreover, the corporation actively provides the public with reliable information on buying, renting, renovating, maintaining housing that supports the valid data for public policy and industry to do their businesses. The main mission of the CMHC is to maintain the decent standard of housing for all Canadians with grants, subsidized rent-units, and mortgage loan insurance to less fortunate income group. In doing so, the Canadian towns and cities are characterized by a socially inclusive housing. This urban character is signified and demonstrated by various types, forms, styles, and sizes of residential, which reflect a variety of income groups in a socioeconomically integrated community. Nevertheless, diverse housing and populations need to be incorporated into the plans that work for economic growth, social stability, and ecological sustainability.

Planning and Transportation

To what extent, planning practice is affiliated with the lifestyle of populations? Ideologically speaking, planning is not impartial from sociocultural values system because it is about people who work, live, and play in its domain. One important factor that is determinant for the way people do in dealing with their habitation, occupation, and recreation is the transportation system. Transportation system has been playing a significant role in the formation, construction, and elaboration of urban settlement. The labyrinth street pattern and form on the hilly sites had been developed by the transportation system based on human, horse, and donkey movement. The birth of ancient gridiron street pattern and form has had something to do with the movement of chariots and warrior parades riding on the horses. In the modern era, the gridiron pattern is important and logical for the development of streets, roads, and utility system because of its practicality for fast construction, efficient operation, and maintenance.

The industrial revolution in Europe and Great Britain had been the milestone of modern transportation system and communication between urban places and regions. The development of mass transportation system, especially with trains and buses, engendered the decentralization of settlements outside the traditional towns and cities at the beginning of the twentieth century. In certain regions of England and Germany, the suburban growth was causing a propensity for the towns to coalesce into mega urban agglomerations or conurbations (Hall 2014, 65). In the United States, the transportation system based on car inspired the architect Frank Lloyd Wright (1869–1959) to develop a concept of the *broadacre city* in 1930-s. Wright's proposal of modern settlements is developed from the with the spread of homes in the network of highways for fast traveling cars; each home is provided with a sufficient area of land for growing crops. The points of connection between groups of homes are the gasoline stations and shopping centers. The similar idea had been proposed previously by Spanish engineer Arturo Soria Y Matta (1844–1929) in 1882 called *El Ciudad Lineal.*

Regarding the possibility of a well-connected network of superhighways, Le Corbusier proposed the idea for *The City of Tomorrow* in 1922 and *The Radiant City* in 1932 consisting of the highly concentrated areas of high-rise buildings in the network of radiant highways. Accordingly, the evolution of transportation system has transformed the way we understand and build the city that is characterized by high-density dwelling towers. Corbusier's human settlement is 'machine for living' that deconstructs the idea of center with the public realm for gathering people. The form and land use of the city is characterized by the perfection of geometric layout that incorporates the mobility and flexibility of people's movement from one place to another. In other words, Corbusier's plan encourage the decentralization of human activities of work, live, and play because of the technological advantage of the transportation system. Corbusier's idea of the city is not the place for walking and cycling but the destination of movements with speed.

Among other modes and means of modern transportation is the use of the private automobile. Since the end of the Second World War, the use of the private automobile has been increasing that has been leading to automobile-dependent lifestyle. Modern urban planning has been always to include the use of an automobile for its spatial structure and capacity. In other words, modern city or town planning is never out of commission for the use of the private automobile. This is understandable because the movement of people is made possible by the network of roads for automobiles. At the beginning of the twentieth century, the use of the private automobile is the manifestation and actualization of individual freedom and mobility; meanwhile, at the end of the same century, a private car is the environmental hazard for climate change regarding its impacts to air pollution and greenhouse emissions. The individual freedom of mobility is not free but costly in terms of operation, maintenance, insurance, and the indirect cost of environmental hazards.

For urbanism, automobile dependence engenders the movement of populations centrifugally from a center. The movement in North America is well known as the phenomenon of the suburban nation. There some reasons for this suburban sprawling movement, such as the increasing land price, the decline of public facilities and utilities, and the deterioration of the environmental quality of the inner-city areas. After the end of the Second World War, automobile industries supplied the market with mass production and affordable cars for middle-class populations; most people in North America enjoyed the economic growth and industrial surplus of various productions, from agriculture and manufacture.

The joy of commuting by private car did not last long. Energy crises in the 1970s alarmed the whole globe that the use of fossil-based energy had the limit and was not sustainable. From these crises, there were some efforts to improve the public transportation system with renewable energy resources. As matter of fact, automobile-dependent lifestyle burdens the road with the inefficient transportation expenditure of energy. The impacts of this automobile dependent lifestyle lead to the inefficiency of land use because parking and circulation areas take up more and more space while the inhabitable and productive areas decrease. In other words, the ratio between the space for car and the area for habitation and production becomes more and more economically inefficient.

Urbanism today and tomorrow needs to be redefined within the framework of safe, healthy, and attractive habitation that engenders economic, ecological, and sociocultural community for sustainable growth. Accordingly, urbanism is necessary to provide the places of interactions that demonstrate the livability in various areas, especially in public places. The livability of places in the urban settlement is indispensable to be supported by a well-connected public transit system. This livability requires pouches and nodes of public realm within the pedestrian network. Moreover, urbanism is characterized by the coverage of public transit service and emergency help; any place in the city is necessary planned and designed within the walking distance of public transit point and the accessible area for evacuation. Nevertheless, the livability of the public realm is necessary to be upheld by the urban ecosystem, which is able to maintain the transformation of energy and materials with low entropy and fewer hazard pollutants.

Planning for the livable and sustainable urbanism is necessary to regard the pedestrian areas that encourage casual social interactions and walking city. Since the human body is biologically designed for physical movements, pedestrian-oriented lifestyle is considered healthy for building communities in the urban area. Moreover, the pedestrian-friendly environment provides people with the opportunity for experiencing the architectural details of the city and the visual interactions with others. All these enrich and enhance the quality of the urban life-world with the ever-changing scenes of the public realms. Nevertheless, the pedestrian-friendly environment is necessary to be supported by a reliable and well-established public transportation system with renewable-based energy. Community planning and transportation planning should work hand in hand to integrate the land use of urban areas into the socioeconomic web of productions, services, and recreations. This implies that the city is not only good for works and businesses but it should be an enjoyable place to live and play.

Moreover, urbanism is not about to build a self-contained settlement but it is to explore and develop a highly concentrated settlement with connectivity to other urban centers. However, each urban center needs to develop its own specific and competitive advantage of the economy and sociocultural that enables them to work together with other urban centers and regions for growing together. In order to achieve this, planning for urbanism is not only to manage the development but it is also to set up the framework of systemic collaborations of various elements, structures, and processes of urban development. In doing so, community planning, regional, environmental, and transportation planning are interwoven and intertwined as an integrated part of the planning system for urbanism. The plans of each component should be synchronized and consolidated within a certain period of time and always open for review, adjustment, and amendment.

Concluding Notes

The nature of planning is imaginatively anticipating in response to local and global needs to be met and issues to be resolved. Formulating the needs and defining the issues are important for setting up the plans. The subject matter of physical planning is the management of resources that enable the local world to establish, develop, and sustain its socioeconomic activities and its ecological processes, land use planning is one of the important aspects of physical planning that manages and regulate the uses of land according to certain criteria for urban livability and sustainability. The official plan is the planning implement at the policy level concerning the strategic direction of urbanism within a five-year period while zoning by law works at technical level how the urban development is managed through the set of regulations, principles, and measurable controls. Nevertheless, all plans are not only the part and tool of development control system but also the essentially integrated structure and mechanism of urbanism. In this point, urban planning should improve the quality of the urban life-world; the goal of planning includes to transform the built environment from automobile dependence to pedestrian-oriented lifestyle.

References

Alexander, Christopher. 1979. *The Timeless Way of Building*. Oxford: Oxford University Press.

———. 1987. *A New Theory of Urban Design*. Oxford: Oxford University Press.

Calthorpe, Peter. 1993. *The Next American Metropolis: Ecology, Community, and the American Dream*. New York: Princeton Architectural Press.

Cullingworth, J. Barry. 1987. *Urban and Regional Planning in Canada*. London and New York: Routledge.

Duany, Andres, Elizabeth Plater-Zyberk, and Jeff Speck. 2000. *Suburban Nation: The Rise of Sprawl and the Decline of the American Dream*. New York: The North Point Press, Farrar, Straus and Giroux.

Duany, Andres, Jeff Speck, and Mike Lydon. 2010. *The Smart Growth Manual*. New York, London, and San Francisco: McGraw-Hill.

Grant, Jill. 2006. *Planning the Good Community: New Urbanism in Theory and Practice*. London and New York: Routledge/Taylor & Francis.

Hall, Peter. 1975. *Urban and Regional Planning*. Newton Abbott, London, and Vancouver: David & Charles.

———. 2001/2007. "The City of Theory." In *The City Reader*, edited by Richard T. LeGates and Frederic Stout, 431–444. London: Routledge.

———. 2014. *Cities of Tomorrow: An Intellectual History of Urban Planning and Design Since 1880*. Hoboken, NJ: John Willey & Sons.

Hodge, Gerad, and David L. A. Gordon. 2008. *Planning Canadian Communities: An Introduction to Principles, Practice, and Participants*. 5th ed. Toronto: Thomson Nelson.

Krier, Leon. 1984. "Houses, Palaces, Cities." Edited by Demetri Porphyrios. Architectural Design 7/8.

Pagliaro, Jennifer. 2017. "Onward and Upward." February 17. http://projects.thestar.com/ontario-municipal-board-reform/onward-upward/.

Thomas, Ren. 2016. *Planning Canada: A Case Study Approach*. Oxford: Oxford University Press.

5

Urban Design and Urbanism

Urban Design as Concept

The theories on urbanism from design perspective have been explored and elaborated by many scholars (Bacon 1967; Alexander 1987; Alexander 1979; Carmona and Tiesdel 2007; Carmona 2003; Lynch 1981; El-Khoury and Robbins 2004/2013; Beatley and Newman 2009), and the important thing of design for urbanism has been seemingly overlooked. Urban design is the way to incorporate the intentionality of urbanism with man-made environment. Normatively speaking, public aesthetics is necessarily represented by the good urban form in the context of social uses that can be expressed in the following dimensions: vitality, sense, fit, access, control, efficiency, and justice (Lynch 1981, 118–19). Still, in terms of public aesthetics, design for urbanism is more than a just technological planning tool. It is ideological in terms of subject matter of power play for political support or rejection. In ancient Greek times, an agonistic way of life was encouraged as part of political games in the search for best (Hooper 1978, 179–92; Thomas 1988, 16). Indeed, in a democratic process, public aesthetics is subject to political discourses and deliberations. Good for urban design is more than just a visually attractive

© The Author(s) 2020   **101**
B. Wiryomartono, *Livability and Sustainability of Urbanism*,
https://doi.org/10.1007/978-981-13-8972-6_5

and appealing form. It needs to be convincing for getting the majority of political support. Even though the definition of public good for forms and meanings is not easily defined (Blau et al. 1983, 300), public aesthetics as urban experience is necessarily related to the concept of civility mentioned earlier. Accordingly, the democratic political process is to diminish the gap of interpretation on good design between professional practitioners and users.

In post-industrial countries, public aesthetics is a battleground of freedom of democratic society and capitalistic free market economy. Here, the quality of the public sphere is marginalized, if it is not colonialized, by commercialization and privatization. The emergence of urban design as a discipline in North America is historically in response to this unique situation prior to early 1980s. It is by and large a critical response to the poor quality of urbanity and cityscape in most North American towns and cities (Carmona 2003, 12). The poor quality is characterized by architecturally disrupted urban form, lost spaces, monotonous urban scenes, and negligent public places. Here, the need for urban design is historically reactive.

Since, as a discipline, urban design is an interface between land use planning and building design, as well as between environmental design and other public works: transportation, utility, and sanitation engineering, its specific function is to build a bridge of inter- and multidisciplinary possibilities of urban development for public aesthetics. Furthermore, urban design is an interdisciplinary three-dimensional planning tool. Accordingly, its practice is indispensably political in interfacing private and public interests as well as their domains. Consequently, it is necessary to work together with: land use, infrastructure, and community planning system in creating, strengthening, reinforcing, and sustaining great and strong urban societies and their places.

By and large, in North America, the adoption of urban design as planning component at the municipal and state policy level has not been in existence before 1970s. Hitherto, there is no precise record available to confirm that urban design has been adopted by their planning acts and implemented in municipalities in North America. Notwithstanding, urban design as public policy is still the work in progress. Some may have been implemented, but in many cases, it is something without precedence in the past. Historically speaking, the original concept of urban design was to replace the notion of civic design (Parfect and Power 1997, 204; Hall 2001/2007). Accordingly, urban design is ideologically a state or municipal intervention for the public good. Historically speaking, everything good with top-down approach is unlikely successful without authoritative forces.

In the world of practice, architecturally detailed design guidance is admittedly acceptable; it provides people with the logic of measurable cause and effect, such as daylight, access, safety, and weather protection. All this is commonly incorporated in site plan application process and review. Thus, the urban design aspect is an integrated part of the development proposal package. Urban design as a policy of local authority needs to be brief and clear, but pregnant in providing goals and frameworks so that development proposals are able to find a common ground on public good (Lang 1994, 109).

In many cases, urbanism by design is a technological instrument and policy tool of development control is implemented in a new development by private developers. Ideologically speaking, it is akin to a privatization of urbanism under private township or subdivision development. Urbanism as a development policy in such private development is practically workable for a new community such as in Seaside Florida (Kolson 2003, 116)—a vacation town designed by Andres Duany and Elizabeth Plater-Zyberk in 1981. Such a private development enables urban design as a development control policy workable, but it is simply a monadic approach, which is regionally isolated from other economic players. How a policy on urbanism could fit in a generic way is probably the crucial question to be addressed. In the world of predominantly existing private enterprises, urbanism is necessarily flexible and modestly conversant in implementing its regulatory principles for public policy (Ratcliffe and Stubbs 2004, 237).

Studies on urban design have been explored and presented by a number of scholars and practitioners from various perspectives and interest. Historical and theoretical aspects and components of urban design have been compiled and studied by several scholars (El-Khoury 2013; Larice and Macdonald 2013; Morris 1972/2013; Hall 2014; Lang 1994; Punter 2009). From practical experience and reflections, a number of scholars and professionals (Alexander 1987; Farr 2008; Brown and Dixon 2014;

Moughtin, et al. 1999/2003; American Planning Association 2006) have made a significant contribution to the founding of urban design as a discipline and area of professional practice. Urban design as a subject with an interdisciplinary outreach owes to the works of Mathew Carmona (2014) and Madanipour (2014).

What is urban design? It is not simply a question of terminology because if we are not able to describe a concept without the thing. Unlike street, urban block, building, park, and square, the urban design does not have a concrete reference. Regarding its nature, urban design is an abstract concept or an intangible thing. However, we know that urban design is about all those associated with making places. Then, we come to the question: What is the thing of urban design? What is at stake in urban design? Beyond all appearances, urban design is a concept. The concept is supposed to have the thing. Otherwise, it is substantially nothing. As a concept, it represents the thing with manifold appearances—what we call phenomena. In everyday life, we encounter the thing in terms of phenomena. We are on the way to find the relationship between the thing and its phenomena that belong to what urban design is.

The light for the question is a clue that urban design has something to do with the built environment. Furthermore, it has something to do with urbanity and development. Inherently, it is also obvious that as a concept, urban design is constituted by two conceptual components: 'urban' and 'design.' It is about phenomena of 'being urbane' by design. What is 'being urbane'? If we rephrase that urban design is a single concept concerning the urban built environment, then we come to realize that urban design is about making something for being urbane. Hitherto, we can see urban design as the process and product of urban development by design. In other words, it is phenomena of design that provides us with scheme and plan of urban development for being *urbane.*

Urban design leads us to find out the relationship between urbanity and making places. This is the search for the opportunity and contribution of design to the process of making places. Under the notion of place, we understand the phenomena of gathering where people, location, and time come into play to designate liveability. The thing of urbanity shows itself in various phenomena of gathering in public space. The inquisitive question

Fig. 5.1 Old Jakarta City Square, Indonesia (Photo by author)

on urban design is how to make public space that creates, establishes, and sustains urbanity.

The relationship is the bond between urban activities in public spaces and the possible form of the urban environment. However, to answer the question properly, we need to think that urbanity is indispensable from liveability and sustainability of living, work, and play in urban areas. Then, we can say that one of the most important things about urban design is about making livable and sustainable public spaces in the city by design. In this respect, the urban design includes designing possible urban form that creates, establishes, and enhances the human experience for work, lives, play, casual encounters, and gatherings in a pedestrian environment.

The built form and the spaces between buildings are the things we deal with in urban design. Form and space are basic components of the built environment. Putting these components into a system of design is what urban design does that works and supports human activities for play, live, and work. Succinctly, urban design is in the process of making urbanity alive from which we experience urban activities as a culture (Fig. 5.1).

Within the urban design system, we will integrate paths, nodes, landmarks, structures, and public transit into a working framework of urbanity. We talk about the urban environment in which the scale, density/intensity, connectivity, diversity, identity, liveability, and sustainability play an important role. Nevertheless, the well-functioning urban environment is not yet perfect without beauty in its form and enjoyable its space. Then, urban design is to create, establish, and sustain something useful and beautiful for urbanity.

Urban Design and Urbanity

The goal of the urban design is to create, establish, and sustain the liveable urbanity based on most of the local characters and resources. Creating and establishing the sense of place are understood as the process of revelation of gathering in which people find their sense of home. It is the sense where people are able to open up themselves to each other based on respect and dignity. Under this notion, people experience that everything mostly their need is in their reach of hand by feet or bike. In this sense, we understand the meaning of places and communities which are not only well functioning in upholding urban culture but also growing up dynamically for beauty. Liveability and sustainability are only possible if our urban environment and communities are healthy socioeconomically and ecologically. Liveability is likely impossible without safe, healthy, and prosperous communities. The task of urban design is to provide an urban environment which is conducive for the liveability of urban activities for work, live, and play in sustainable manners.

Practically speaking, the thing of urban design is a public realm for urban activities. Urban design is not only to resolve the conflicts of interest and use in the public realm but also to bring about liveability and sustainability of urbanity. Creating and developing public realm in Canadian urban context are barely workable without being able to unveil pedestrian-oriented neighborhood. This implies to turn out the North American lifestyle from the individual mobility with the car to public transit. Consequently, urban development is not only to deal with physically restructuring urban development, but it is a philosophical and political transformation at the individual and community level. Theoretically speaking, it concerns with the way people think of what urbanity is. On the other hand, it is also about political will concerning a sustainable economy and environment. Even though it is not an easy matter, hope and direction for the viable option are apparent. The problem is that the process requires periods of time and community engagement. In addition, the transformation process needs a programmatic implementation that includes a socialization and political engagement of the stakeholders concerned. Public meetings and workshops are parts of this urban development process.

One important goal of the urban design is to create and develop a scheme and plan for the livable public realm. Under the notion of the public realm, people come to terms with casual encounters with each other with safe and pleasant amenities as well as the attractive built environment. The public realm is characterized by the sense of destination of their movement and civilities. In short, urban design is the art and science of making public places for the urban community. In the Canadian context, public realms take place potentially at the intersection of streets and sidewalks along the main street. Needless to say, there are some issues, opportunities, and constraints that need to be worked out by design, such as the proportion of street width and building height for pedestrian scale, urban block structure, building density, streetscape architectural character, environmental safety, and public supportive environment.

The vital idea of urban design is to integrate the public realm as the outdoor living space of urban community. The integration includes incorporating public transit system and green conservation system within the urban design framework. In doing so, urban design is inseparable from sustainable development. Buildings, open spaces, and streets are the grains or elements of urban design. Bringing together buildings and streets as well as open spaces is the way urban design works. However, bringing together elements without strong structure, clear canon, and principle will lead to nothing but chaos. We need the ordering system as an urban design tool in terms of codes, regulations, and guidelines. The practice of urbanism needs urban design principles to create, ensure, and sustain the liveability of the public realm. Beyond representations and rules is the semantic system of urbanity. The semantic system gives urban design goals and meanings. The sense of urbanity arises from this system.

Urban Design and Codes

The necessity for codes is to control the principles of appearance and the technical standards for the healthy and safe environment. Building codes have been working as the set of rules and regulations for technical standards for building construction since the beginning of the twentieth century. However, the architectural character and visual quality of the built environment are not managed and covered by building codes. This provides either the room of creativity or the indifferent loophole for architectural expression and articulation. Since the quality of urban livability is possible because its built environment architecturally and economically attracts people to be there, the architecture of place plays an important role in urbanism. So that, there is the necessity for creating and establishing the built environment of a certain location with strong architectural characteristics and ambiance. Prior to the 1970s, he dulls and unvarying streetscape and architecture of urban settlement in North America are degenerated by the zoning regulation; this set of rules is in favor of the land use policy based on a single function.

As a set of rules of urban development, the zoning by-law—in many cases—is not well provided with a three-dimensional objective concerning the architectural and social quality of the built environment; there is mostly no clear connection between rules and physical outcome (Talen 2012, 5). The architectural codes are the system of planning and design that transform design ideas and principles into actual form according to public interest and preference. Even though codes seem the set of generic and common principles, in its application it is necessary to be specific and unique for certain location and context. This is in order to avoid the monochromatic and unvarying streetscape and characterless built environment. From the interest of public good, local community needs the access to participate to craft and develop their built environment. Such participation today becomes a community need for their sense of belonging to the place where they live, work, and play. Urban architectural codes are necessarily the outcome of the local community's commitment and vision. Without such a community participation in the planning and design process, codes remain ineffective as a planning tool for building healthy, safe, and lively built environment with vibrant public places.

As mentioned above, public interest and preference are essential for land development and redevelopment in urban areas. The public good is necessarily understood as the basic condition for safe, healthy, and lively built environment for all. This condition is necessarily enhanced and enforced with the architecturally appealing built environment that is conducive for circulating and informal gatherings of people. Urban design codes make this condition possible provided that all stakeholders in the location have been involved in their planning and legislature process (Dobbins 2011). Planners and designers from private and public sectors have to listen to the community's need and vision; this does not mean that the professional leadership is weakened and diminished. On the contrary, planning and design expertise leads the whole process of crafting and developing the codes based on what is necessary and potential for local resources. Such a leadership is neither patronizing nor directing but unveiling and divulging local potentiality toward the excellence.

Several efforts and endeavors have been made to amend and furnish the zoning by-law with building codes and other technical requirements for public safety and health. In many cases, zoning discourages mixed uses in urban development. The need for urban design is primary not because zoning by-law does not work well. The need for design principles is obviously from the exigency of a better quality of the built environment for livability and sustainability. The design is considered a potential way for the betterment of urban environment; the problem for this betterment lies in the planning process how design principles become applicable architectural codes. Design principles need to be supported by studies and best practices. In doing so, all design principles are workable at two levels of implementation: generic and specific location. Urban design codes are not intended to create and establish a certain architectural style and microexpression of the built environment. The codes should be able to manage generic scale and proportion of form and space such as for streetscape section in a certain location, but not to advocate uniformity of scale and proportion regardless of local specific condition.

Skeptic responses to the implementation of form-based codes include the view that such rules restrict creative and innovative architectural expression. The other responses oppose the implementation of such codes because of its deterministic nature on behalf of aesthetics. The opposition becomes stronger if the planning process concerning the implementation of codes does not work democratically based on community participatory approach. Nevertheless, codes are the set of planning and design tools that enable people to identify themselves with their place so that they feel at home as a member of community and location. From this point, codes are inseparable from the community's need for the mnemonic identity of place and for the sense of belonging together. In other words, codes should not simply work as technical planning and design tools, but work as social codes of the local community that makes them proud of their place and tradition.

Urban Design as Public Policy

Urban design as public policy has been extensively studied, explored, and presented by a number of scholars (Barnett 1974; Punter 2007; Tiesdel and Adams 2011). Urban design is barely to be a public policy without the support of all stakeholders in urban development. In democratic countries and states, being aware of the dominant role of the private sector in urban development is one important point for successful urban design as public policy. In North America, the commitment to privatize domains has been long-standing flesh and bone. Here, the tension between the tendency toward autonomous individualism and communitarian citizenship is in the scene of urban design playground. It is not only about urbanism and sub-urbanism but also about the whole concept of what home is. For example, Canadians today are not looking for 1960s' Los Angeles for their idea of home. However, most private corporations and entrepreneurship in the Canadian context are apt to less government and prefer to private initiatives and freedom; meanwhile, a community for public citizenship puts civic welfare in their favor. Suburbanism in the Canadian context is more about natural living with a family-oriented system of values. In dealing with this condition, urban design plays an important role in carrying out the differences in a win–win situation with tangible options. Of course, it is not only interfacing the public and private realm by design but also transforming urban areas for a family-friendly environment.

Moreover, urban design as public policy is not simply about planning and procedure matter. Competition of investment between metropolitan centers and suburban areas is apparent and evident. Making urban design principles as public policy will be a challenge for such competitive investment. This is usually related to the gentrification of declining urban areas by private sectors for commercial, residential, and industrial uses. Urban design as a strategic and practical tool for planning system plays a significant function for towns and cities that attract potential investment for urban redevelopment, rejuvenation, and revitalization. Nevertheless, the function of urban design scheme and policy is necessary to potential investors for the return on their investment. Great places by design in a capitalist economic system are not impartial from the profit-making of big corporations. To a certain extent, the dynamics of financial power and regulating authority often come into hard negotiations. The problem of making great places in North America is that development policy is not always in concert with the business trend and population growth. Conflicts of ideas and problem-solving options often happen between the community and private developers. Most common issues are about the conservation of social historical concern and business feasibility. If the new development's initiative and public interest are not able to come to terms, both parties should go to a tribunal or court. In this case, the question of the sense and essence of development remains: What is a good urban design for?

Normatively speaking, the necessity of urban design as public policy is to bring the awareness of urban citizens for the improvement of architectural quality for their place of work, live, and play. In order to achieve this awareness, the access to participation in the planning and design process should be provided by the law. First of all is to achieve a collective vision and agreement concerning what the public good is. By having a clear vision of development, urban design principles can be discussed, debated, deliberated, and resumed for a further step in the planning and legislature process. Nevertheless, the set of principles must be sensible and reliable in carrying out conflicts of interest into the agreement. In other words, sustainable urbanism as development vision is not simply a matter of regulations or technical options, but a framework for collaboration, participation, and resolution. It is the outcome of the extensive process of community engagements and market economic studies.

In reality, the challenges for public policymakers are not merely legal, formal, and technical matters, but also to set forward the agenda of urbanity for large public interest in a delicate manner. The most potential drive for the agenda is an investment in various buildings. Albeit formal directions for public good remain the guiding principles, policy playground needs flexible options in accommodating investment as well as a community concern. Urban design is one of the flexible options. Design in this sense is a policy tool that brings about conflicts resolution intangible options. What can be offered by design is a range of strategies to the technically specific implementation of development policy. This includes issues concerning: diversity and multiplicity of uses, connectivity, and linkage of transit, building density, built form and streetscape, place character and identity.

Density and Intensity of Uses

The density of population and building intensity are the factors that enable the urban environment to find its physical character. The concentration of activities for work, play, and life is one important component of density and intensity which are expressed in building a form with multi-stories structure and large size volumes. The density of the urban area is indicated by the number of populations and units of dwelling, whereas intensity is to figure out the proportion of the built-up area to the site. Intensity shows the relationship between building and site that gives us the idea of areas covered with buildings and open spaces. Building coverage in the most urbanized area is about between 60 and 80%. The other indicator of intensity is the floor index or floor area ratio mentioning the total areas of the building in relation to the area of the site. From this floor index ratio, we are able to draw the number of floors acceptable and suitable regulated in zoning law. In other words, density and intensity are indicators of urbanity shown with compact urban form. Nevertheless, the problem of urban design is not only to create a compact urban form, but also to design and treat the space between buildings to be useful and beautiful.

Making policy on density and intensity is supposedly not reactive, ad hoc, disjointed, incrementalist, and opportunistic. Instead, making policy for urban density will not make any sense without a solid ground on the value of land and development. Behind the scene and down the road, the market economy of the location plays a very important role in determining the success of the development. However, a comprehensive development policy for infrastructure that supports certain density is necessarily provided, especially for the public transit system and its network.

In most urban areas, the policy on the density of populations and building intensity is challenging in two ways. Firstly, increasing density for public transit ridership will require restructuring urban blocks with a supportive land policy for live/work uses. This will lead to a possible transformation for the new streetscape and new urban scene for betterment. Increasing density of a location means also keeping up its current district policy in compliance with the whole new development policy. Secondly, policy for increasing density and intensity has to deal with the increasing capacity of road or street. Consequently, slowing down traffic becomes necessary. When intensifying the land use is implemented, building a subway or other mass-rapid transit system becomes an option for such transformations.

Furthermore, what urban design can offer is to offer the positive image of locations with various development scenarios. This includes to unveil local resources and characteristics as well as to open new socioeconomic opportunities. Urban design can be an effectively appealing development tool in unveiling in a proactive manner as well as in exploring the potential urbanity of a location concerned. Urban design plays a significant role in development projects because of its intuitive approach. Regarding its intuitive method, urban design is the only tool of urban development that can defy empirical and rational approach in the making of urbanity. The unique capacity of urban design is to demonstrate the possibilities of development in intangible manner. So far, urban design is the best tool to demonstrate the value of the density of populations and building intensity. Nonetheless, density is barely implemented without diversity of activities because its aim is to increase and improve the interactions and collaborations among members of the community.

The Diversity of Uses and Forms

Diversity includes a variety of activities, forms, styles, and populations that proffer the dynamic of liveability. The importance of diversity lies in its dynamic potency for various interactions and attractions as well as occasions. Socioeconomic diversity is the basic condition of the economically sustainable community. In the natural environment, biodiversity enhances the mutual interdependence of the life cycle in the web of food and energy. The main contribution of diversity in the natural ecosystem is to keep the entropy low because the system has fewer wastes and by-products; everyone's output becomes another one's input. As a matter of fact, diversity broadens the role and function of elements in a system for collaboration and mutual synergy. With regard to the diversity of community and ecosystem, urban design will accommodate the possible condition of liveability of the urban environment with active options.

In its implementation, diversity in urban design is workable in developing mixed-use activities. In many cases, mixed-use activities engender rich opportunities and possibilities of collaboration in the community. Mixed uses are the foundation of liveable diversity that establishes and sustains vibrant activities provided that the activities are socially and economically supported and surrounded by enough healthy residential units. However, the diverse uses do not generate interdependency for growth and sustainability when the populations there are mostly unemployed. With this regard, urban design is not about to make a good scheme but also to work together with the economic feasibility of the location concerned.

Moreover, diversity is able to multiply their interdependency of activities for productions and services. This happens as a consequence of compact design with a well-integrated access to infrastructure neighborhood. Compact design enables efficient interactions for social and economic activities. This is commonly accessible within five to ten minutes walking distance, within 400–600 meters radius. Historically speaking, the good relationship between compact urban form and diverse uses has been achieved by the main street of village towns in North America, High Streets, or marketplaces in England. The more diverse is the concentration of activities, and the more vibrant are the streets or places. Besides, the generator of multiplying activities is actually the mixed use of land with ample opportunities for intensification.

Economically speaking, small businesses with a wide range of diversity play an important role in the successful traditional main street activities. In most traditional towns, the function of urban design is to enhance and strengthen local identity. However, the integration of transit hub into the areas sustains and enhances the livability of urbanity. In the capitalist economy today, such diverse uses and retails have been integrated by private developers into a large scale of mixed-use development or big-box development. Since 1970s in North America and elsewhere in the globe, privatization of common elements of work, live, and play has been being taken over traditional public realm with diverse uses along the main street.

Connectivity, Linkage, and Transit

Connectivity of people's movements is the key to the successful urbanity design. Connecting all building entries, open green spaces, points of interest into a pedestrian, and bike cycle network is one urban design step to set up its essential framework. The purpose of this connectivity is to ensure that the built environment is pedestrian friendly for direction, orientation, accessibility, safety, and convenience. Nevertheless, connectivity is by no means a traffic-through environment for automobiles. Instead, urban design is to create a public traffic supportive community and to reduce private cars on the streets.

Furthermore, connectivity by design is about giving various options and opportunities for movement and destination that have convenient access to retails and workplaces. Under the notion of connectivity, urban activities are brought together not only in a diverse and well-distributed way but they are well designed to ensure public safety and a healthy environment. In doing so, all spots are in a well-organized system in compliance with the safe and healthy standards of urbanity for production, leisure, transit, maintenance, and service. Nevertheless, connectivity by design is necessarily thought to ensure that business and leisure in the location are potentially improved and attractive.

Slowing down traffic on the arterial roads for major cities in North America is a challenging option because it costs traveling time between two destinations. The other viable choice is to direct the commuters to the nearest highway. The problematic case is for a district area at the intersection between major intersections. It is problematic because such a street has been the line of various business and commercial activities. Transforming the land use of linear to highly concentrated form will affect the economic value of the land. This transformation is challenging in most towns and cities because this will transform the lifestyle from automobile-dependent way of life to reliant on public transit.

Improving connectivity is making all potentially vibrant urban areas in an efficient nexus system, in which urban design is incorporated as public policy for green energy use and attractive corridor. Connectivity is about connecting the dots in the nexus of public transit that offers serial urban scenes for efficient transportation and enjoyable riding. Urban design is at capacity for improving connectivity by designing transit stations and their junctions with unique and attractive forms and places. Integration of various modes of transportations as well as uses is commonly practiced for new transit hubs. From business perspective, a partnership of private and public sector is workable with various ways, such as build operation transfer, concessions, design build operate, joint ventures, partial divestitures, and public assets full divestiture schemes. It is commonly associated with a transit-oriented development that will create and generate new activities with the environment-friendly facility.

Built Form; Scale, Proportion, and Massing

Eye-catching streetscape and remarkable built form play an important character in creating and establishing the sense of place. This includes distinctive architectural style, the use of locally recognized material and form, as well as the enhancement of historically existing iconic buildings and landmarks. Urban design is able to create remarkable and memorable places that make their people as community proud of. This builds an emotional relationship between people and their place in terms of home. Beyond its technical options, urban design as public policy is to concern with this issue. A strong community is barely achievable without the sense of place they call as home. By design, a strong community is workable with creating places where people are able to interact and know each other by faces. Pedestrian-friendly places with various amenities are conducive to building a strong community.

The urban scale that urban design deals with is to achieve the proportion between the width and height of public space between buildings, in such a way where people find themselves in proximity with each other. This is the scale where the sense of enclosure and openness comes into play with the flow of pedestrian and bike cycle networks. Urbanism is not necessarily to build gigantic and large buildings where human dimension becomes to be indifferent. What is really matter for urbanism by design is to achieve well-proportioned scales that people are able to relate their human scale to parts or whole building without disorientation and confusion. The distance to recognize blocks, patterns, and units by naked eyes has the limit for the convenience of what we call human scale. On the other hand, the ground and street level is the basis of the urban dimension to experience the daily urbanity. Under this notion, urban design is about making our walkthrough experience on street level convenient, pleasant, and open for visual surprises. Treating urban elements in convenient relation to pedestrian level is one important option to deal with urban pedestrian scale. In doing so, the proportion of enclosures and open spaces is necessary to be in compliance with human scale for proximity.

The notion of human scale is to relate everything objective for our eyes in a recognizable and familiar dimension to our body. This implies that buildings and open spaces need patterns and units in relation to the human dimension. Eye level on the street is the best common ground for streetscape experience. Walking through on foot is also important to measure to experience human scale. Two-story buildings stand face to face on the 12–15 meters right of way street space are mostly acceptable for pedestrian convenience; as rule of thumb, commonly found in historic towns, the proportion between the height of both side buildings and the width of the street is one to 1: 2 or 1: 3. Maintaining the streetscape of traditional townships is usually challenged by more capacity for business opportunities. Even though such traditional building structure is acceptable for pedestrian scale, their space is less favorable for new investment. Restructuring urban blocks and their building density is an option but the historic character and identity of the place should be not at stake of new development.

As mentioned earlier, urban design is to bring about forms and places within the scope of the human scale. This goal is achievable not only by the creation and establishment of urban edges but also by urban landscape elements like trees, street furniture, fixtures, utilities, and public arts. Nevertheless, without having a story, experiencing urbanity is not quite perfect. Built form and its places are signs of historical traces and memorable constituting structure of such a story. Urban design is challenged to set up the plot and scheme that enable people to read and build their stories about the places they visited. For urban design, built forms and places are not only necessarily attractive and appealing but they are also to represent locality. In doing so, preserving and conserving historical and local resources are inseparable from keeping the cultural relationship between people and their place.

Identity & Character

Urban form in an urban design context is to bring all buildings in the area as a whole system or complex. This implies that urban design is to put together units and elements of the built form into an aesthetically harmonized composition. The use of repeatable elements, forms, patterns, styles, and details is commonly practiced for such a composition. However, a clear and strong structure or framework is an important aspect of design so that the interplay between forms and domains does not lead people into confusion and disorientation, especially in finding emergency exits, directions, and destinations. Moreover, a good urban design is an architecturally integrated composition with clear characteristics. In order to achieve this, it is necessary to design buildings in the areas into two categories: architecturally defining edges and iconic structures. The defining edge buildings are parts of the making public spaces; these buildings define and construct definite places for movements, casual interactions, and plays. Meanwhile, iconic buildings provide people with landmarks of orientation and navigation.

Architecturally speaking, the edge buildings act as street walls that define enclosures, squares, and corridors in various scale and volume. Defining street edges by buildings and landscape elements is indispensable from the street pattern and open space structure. From this point, defining urban edges is not only to shape urban form with street walls. It is about defining public spaces in a structural system of corridors and nodes. A well-structured system of urban edges creates and opens up an attractive interplay of options for direction and openness.

The possible urban form is necessarily treated in a whole system that consists of various components but they show common marks or characteristics that they belong together. In such unity, every component or building is unique and visually recognizable as part of a whole arrangement. This uniqueness is a kind of uniformity but without losing variety and particularity of each individual building. Unique elements include corner buildings, entrances, facades or fenestration, and other details for signage and lighting. All of these need special treatments that bring all these into visually interactive at the pedestrian level. At this human scale, every detail and particular elements are potential to play an important role in the enhancement of urban experience.

The other important aspect of urban form is to give people clear visual direction and orientation as well as the enjoyment of movement. Repetition and modulation of similar elements and patterns on the facade are commonly applied to accompany people's movements on the sidewalks of the street. The urban form should make sure that such movements are continuous and smooth toward destinations. Permeability and easement from and to each other places are necessarily integrated into the urban form design. In doing so, human movement is not strictly confined by the bulk of building masses and open spaces. Short linkage and variety of routes make the pedestrian environment enjoyable and convenient.

Street furniture proffers people to have an option for rest on benches for enjoying the scenes in public spaces. Urban experience in urban design is a potential event in various scenes that take place in urban public places as their stage. Urban experience consists of various scenes that are able to build an urban story for everyone. Stories about a city are potentially

Fig. 5.2 Entrance to Deoksugung Museum of Arts, Sejong-daero, Myeong-dong, Jung-gu, Seoul, South Korea (Photo by author)

inscribed on well-designed places and urban landmarks. Such structures and places are at capacity to represent and incorporate local values and memories. The role and function of urban design are to pitch such stories in urban structures and places. In doing so, urban experience is not only strolling along its sidewalks and places but also getting familiar with locality and its urbanity; each locality has its own urbanity in terms of local cultures and prides. The successful urban design is able to unveil local urbanity in its artifacts and places. Such structures and domains are unique and constitutive to build collective memories of its residents and visitors (Fig. 5.2).

Concluding Notes

Urban design as public policy is not only to set forth technical options for urbanity. The significant capacity of urban design is to explore potential opportunities based on intuitive and artistic vision. This role and function will work for hand in hand with the empiric and rational approach to planning in a complementary way. At the strategic level, urban design as public policy is about making the vision of urbanity. In creating and establishing such visions, designers are to deal with the options for bringing about conflict of interests into viable and acceptable solutions. In regard to public policy, designers are not dreamers but listeners who put every idea and thought into a workable scheme and plan for a sustainable future. Regarding its capitalist economy, urbanism in North America is viable and workable, provided that its approach and framework are economically feasible for the return of investment, or at least potential and prospective for economically generating opportunities. In capitalistic and advanced industrial worlds like in North America, urbanism by design as public policy is necessarily open for locally and partially customized innovations. Today, drafting urbanism as public policy is necessary to dismantle all teleological premises and hope for social order and community rhetoric of sustainability. The era of the grand scheme is over. Current urbanism is to welcome to time-consuming deliberations and public meetings for a plan or design layouts. The real problem of public aesthetics is a continuous public process and open discourse on the public good as an open-ended system.

By nature, state planning is political intervention with two missions: as a regulatory and administrative tool. Ideologically, planning is to protect, maintain, and sustain investment. In the course of the history of North American urbanism, planning is a strategic instrument of political rationalization for capital accumulation and circulation. The barometer of success for urbanity has been mostly indicated by economic growth, revenue increase, and employment rate. Less attention has been given to measuring up the performance of public services and the quality of urban places. Urbanism by design is considered pivotal for figuring out cityscape and for providing people with places for the urban experience. A well-designed public realm is an outcome of the well-organized policy of urbanism. And good urban design represents urban culture in terms of attractive and pleasant built form. How can we know the quality of urbanity without experiencing its public realm?

Indeed, the necessity for urbanism is never free from the exigency of a partnership between the private and public sectors. The partnership is not only economically imperative but also environmentally crucial. It is the main chance and challenge for any initiative of urban design proposals. This includes putting all options on the table for discussions and negotiations. Needless to say, the predominance of the private sector is evident. What the public sector could offer is to mobilize expertise and communities in formulating public good. The logic of capitalistic commodity and the imperative of moral for sustainable future need reconciliation at any level of interests, from ideological to the practical platform.

References

Alexander, Christopher. 1979. *The Timeless Way of Building*. Oxford: Oxford University Press.

———. 1987. *A New Theory of Urban Design*. Oxford: Oxford University Press.

American Planning Association. 2006. *Planning and Urban Design Standards*. Hoboken, NJ: Wiley.

Bacon, Edmund. 1967. *Design of Cities*. London: Thames & Hudson.

Barnett, Jonathan. 1974. *Urban Design as Public Policy: Practical Methods for Improving Cities*. New York: Architectural Record Books.

Beatley, Timothy, and Peter Newman. 2009. *Green Urbanism Down Under: Learning from Sustainable Communities in Australia*. Washington, DC: Island Press.

Blau, Judith R., Mark La Gory, and John Pipkin. 1983. *Professionals and Urban Form*. Buffalo: SUNY Press.

Brown, Lance Jay, and David Dixon. 2014. *Urban Design for an Urban Century: Shaping More Livable, Equitable, and Resilient Cities*. Hoboken: Willey.

Carmona, Matthew. 2003. *Public Places, Urban Spaces: The Dimensions of Urban Design*. London: Architectural Press.

Carmona, Matthew. 2014. *Explorations in Urban Design: An Urban Design Research Primer*. Surrey: Ashgate.

Carmona, Matthew, and Steven Tiesdell. 2007. *Urban Design Reader*. London: Architectural Press.

Dobbins, Michael. 2011. *Urban Design and People*. Hoboken, NJ: Willey.

El-Khoury, Rodolphe. 2013. *Shaping the City: Studies in History, Theory and Urban Design*. London and New York: Routledge.

El-Khoury, Rodolphe, and Edwards Robbins. 2004/2013. *Shaping the City: Studies in History, Theory, and Urban Design*. London: Routledge.

Farr, Douglas. 2008. *Sustainable Urbanism: Urban Design with Nature*. Hoboken, NJ: John Willey.

Hall, Peter. 2001/2007. "The City of Theory." In *The City Reader*, edited by Richard T. LeGates and Frederic Stout, 431–444. London: Routledge.

———. 2014. *Cities of Tomorrow: An Intellectual History of Urban Planning and Design Since 1880*. Hoboken, NJ: John Willey & Sons.

Hooper, Finley. 1978. *Greek Realities, Life and Thought in Ancient Greece*. Detroit: Wayne State University Press.

Kolson, Kenneth. 2003. *Big Plans: The Allure and Folly of Urban Design*. Baltimore: Johns Hopkins University Press.

Lang, Jon. 1994. *Urban Design: The American Experience*. Hoboken, NJ: Wiley.

Larice, Michael, and Elizabeth Macdonald. 2013. *The Urban Design Reader*. London and New York: Routledge.

Lynch, Kevin. 1981. *A Theory of Good City Form*. Cambridge: MIT Press.

Madanipour, Ali. 2014. *Urban Design, Space and Society*. London and New York: Palgrave Macmillan.

Morris, A. E. J. 1972/2013. *History of Urban Form Before the Industrial Revolution*. London and New York: Routledge.

Moughtin, Cliff, Rafael Cuesta, Christine Sarris, and Paola Signoretta. 1999/2003. *Urban Design: Method and Techniques*. Oxford and London: Architectural Press.

Parfect, Michael, and Gordon Power. 1997. *Planning for Urban Quality: Urban Design in Towns and Cities*. London: Routledge.

Punter, John. 2009. *Urban Design and the British Urban Renaissance*. London and New York: Routledge.

Punter, J. V. 2007. "Urban Design as Public Policy: Best Practice Principles for Design Review and Development Management." *Journal of Urban Design* 12 (2): 167–202.

Ratcliffe, John, and Michael Stubbs. 2004. Urban Planning and Real Estate Development, Volume 8 of the natural and built environment series. London: Taylor & Francis.

Talen, Emily. 2012. *City Rules: How Regulations Affect Urban Form*. Washington, DC: Island Press.

Thomas, Carol G. 1988. *Path to Ancient Greece*. Leiden: Brill.

Tiesdel, Steve, and David Adams. 2011. "Real Estate Development, Urban Design and the Tools Approach to Public Policy." In *Urban Design in the Real Estate Development Process*, edited by Steve Tiesdel and David Adams, 1–14. Oxford: Willey Blackwell.

6

Environmentally Friendly Urbanism

Sustainable Urbanism

There are several publications concerning sustainability in relation to architecture and urban development that their findings have inspired and supported this chapter. The works include the guidelines and case studies (Steele 1997; Thomas 2003; McLennan 2004; Riddle 2004; Hester 2006; Friedman 2006; Farr 2008) and ideas and options toward sustainability (Newman 1996; Hester 2006; Beatley and Newman 2009). Undoubtedly, the publications mentioned above discuss and survey sustainability in the field of planning and design which are significantly useful for academic theory and professional practice. While these scholars were largely marvelous in their reviews and propositions on sustainability, however, there has been always room for this chapter to sum up and set up a theoretical framework for sustainable urbanism. All this is considered necessary so that urbanism has a conceptual framework of sustainability by planning and design. Such a framework is necessarily holistic but technologically viable and institutionally workable. The complexity and contextuality of urbanism should be overcome by strong principles with flexible adjustment.

© The Author(s) 2020
B. Wiryomartono, *Livability and Sustainability of Urbanism*,
https://doi.org/10.1007/978-981-13-8972-6_6

Lack of systematic treatment of sustainable knowledge for planning and design practice is not the main reason for writing this chapter. Rather, this study is more interested in finding the very fundamentals of sustainability for urbanism. The focus of this inquiry is actually to fill the gap between theory and practice of urban livability and sustainability for public policy that guides and directs urban development with respect to land use and urban design discipline.

Generally noticed that sustainability has something to do with: the use of organic materials, renewable energy, and environmental conservation—by reducing consumptions, pollutant emissions, and wastes. Indeed, all this is functionally not enough for urbanism. The core of vital urbanism lies in its healthy and vital economy (Bairoch 1991). Environmental sustainability is likely less useful without a synergic scheme with economic growth and public safety. Integrating environment and economy have been intensively discussed by Gouldson and Roberts (2000). Accordingly, new institutional reorganizations and integrated planning instruments are necessarily created and developed. Indeed, in order to put sustainable policy for urbanism in the context of the operational synergy of the environment, sense of community, public safety, and economy, a comprehensive approach is necessary. This study is in the search for the depths of such an approach. It is presented neither as a new theory nor as a critical review of available theories on sustainable urban design. Rather, it is considered a tentative inquiry based on practical experience in the field of urban design and land use planning.

By nature, the study is a descriptive analysis in order to unveil, dismantle, and unfold sustainable aspects of the urban environment. The purpose of the study is an attempt to bring forth and to constitute a theoretical framework of sustainable urban design. In doing so, the study is expected to be useful for and contributive to the knowledge and practice of land use planning and urban design. It is commonly argued that a high standard of environmental policy and public safety is necessarily in compliance with affordable economic development. And it does not make any sense promoting environmental protection without encouraging economic growth.

Theoretical Context of Sustainability

The integration of environmental and economic sustainability needs to establish a theoretical framework of sustainable urbanism. Since the development of the built environment is physically technological and socially institutional, policy on sustainable urbanism needs to address and rely on the integrations of innovative technology and adaptive institution that viably work for anticipatory options—instead of merely reactive planning and design solutions. Green technological modernization is, of course, not the only approach toward sustainability. It might be compelled by political leadership and public awareness of the necessity for environmentally friendly urbanism.

To certain extents in North American context, this includes the integration and accommodation of environmental challenges into a capitalistic model of development. Then, contestations of policies between environmental and economic interests come into enigma or in the state of never-ending debates. The problem of sustainability is unavoidable from the transformations within societies. In capitalist Western civilization, the power of such transformations is impossible without the role and support of private sectors whose interest is obviously in gaining profit margin as wide as they can. On the public part, regional governments and municipalities remain mostly in reactive accommodative or passive defensive position in response to development proposals. Obviously, an accommodative platform for collaborations and partnerships is in need of sustainable urbanism. Since changing minds is required in dealing with sustainability, democratic platform and process are necessary. All of this is inescapable to deal with the question of why sustainability? It is not simply a question concerning environmental protection. Rather, it is considered a multidimensional question of the life and death of urbanism. Since most of our problems with sustainability are inevitably technological, the answer to the question should be searched and found within the technological system as well.

In developing a conceptual model of sustainable urbanism, land use and urban design policy should create and incorporate technological innovations and its instrumental institutions within a democratic and sustainable system of capitalist society. Planning and design should be seen as a technological approach to urbanism that concern on the quality of the built environment. In order to achieve sustainable urbanism, such a technological approach for built form should be fundamentally comprehensive and deeply rooted in the necessity for the cultivation of great urban quality of life.

Conceptually, developing such a theoretical framework is not necessary from scratch. Kevin Lynch (1981) has set out the foundation for exploring the city form in accordance with aesthetic and environmental principles that comprise planning, functional and normative theory. In terms of theory and practice, Hester (2006) provides us with explorative study in bringing together applied ecology and participatory democracy with concrete examples. Again, sustainable urbanism has been underway. Our contributing part in this study is neither a compilation of aspects and categories of sustainability nor an advocacy of sustainable politics on urbanism, but simply the search for planning and design instruments for public policy. In drafting the policy of sustainable urbanism, it is considerably important to answer the questions why we need sustainability and what are the structures and components that constitute its conceptual existence. Nevertheless, prior to all efforts, changing minds for sustainability is noticeably more important and strategic than changing the environment.

The necessity for sustainability is functionally and normatively indivisible ground with the goal of planning and design. Functionally, sustainability promises the healthy environment for generations. In order to keep this promise, codes and norms are necessarily incorporated as cultural values that are able to shape and enhance future societies with environmental and social responsibility. By doing so, the need for sustainability is not simply moral. Rather, it is practically imperative. Otherwise, the globe will be overwhelmed by inextricable planetary crises and collective self-destruction.

Theoretically, sustainable planning and design in a narrow sense is an ecological scheme or project that enables us to preserve and conserve the use of land, energy, and resources indefinitely. In the broadest urban context, the use of land, energy, and resources includes activities of work, live, and play which are not limited only in terms of physical environment, but also that of the economy and social life. Practically, sustainability in the built environmental design is a system and process for making a better place with less entropy by using regenerative energy and organic materials. The main purpose of sustainable land use plan and urban design is to bring forth, develop, and establish a scheme of a built environmental system that works for: energy conservation, economic growth, and healthy environment, as well as safe and attractive public space. How to achieve this sustainable scheme?

In order to answer the questions, it is necessary to consider the aspects and principles that work within a natural ecosystem for sustainability. Ecologically sustainable in nature is the state of being in which beings are in the flow of energy transformations. In practice, the natural processes work for sustainability with various phenomena of succession, evolution, extinction, and colonization. The problem is how such processes are compatible and transferable in the man-made environment. First of all, it is worth noticing that in the built environment transformations are not simply about energy in terms of the food chain. Rather, changes and alterations happen and work naturally and economically. Nature and human system are in interactions and exchanges which are not merely about energy, but also about all resources, from matter to money, as well as from knowledge to skills. However, nature has been giving us valuable lessons that sustainability works within the phenomena of diversity, connectivity, functionality, mutuality, perpetuity, and recoverability (Beatley and Newman 2009, 200; Jerke et al. 2008, 160, 230). All these are considered as technological instruments of urban development for a sustainable economy and environment.

Variability

In nature, variety as well as diversity (Shachak et al. 2005, 301) is functional and essential for ecosystem consisting of a variety of species, genre, classes, types, groups, and sorts. Since living in nature is interdependent interactions and shares of space and time in the same habitat, diversity is the necessary condition for ecosystem so that every organism is able to live with its particular role and function in the circulations and distributions of the nutrient system. In the human system, the importance of diversity is not simply for the natural environment, but also for the socioeconomic and political life of the community. Jan Jacobs prescribes diversity in urban areas with the concentration of people as the necessity of urbanism that underpins the livability of the public realm (Jacobs 1961).

Since diversity is about a variety of activities, uses, forms, styles, communities, and populations that work together for the dynamic liveability, elements and components of the system have to have mechanisms and processes of evolution, succession, and regeneration. In nature, such mechanisms and processes are self-regulated in terms of homeostasis. However, in the human system, it is neither easy nor instant. Rather, the human system needs cultural endeavors in order to defy greed and domination; therefore, diversity should be understood within unity in the democratic process (Hester 2006, 197).

Moreover, the phenomenon of diversity stipulates the necessity for a community consisting of various backgrounds, occupations, organizations, educations, ages, affiliations, and associations. The sense of diversity lies in its potential interactions for collaboration and cooperation. Nonetheless, the diverse system enables its components and elements learning to live, work, and play together in terms of community. Diversity in nature is a necessity for engendering any element and component to play their particular and specific role and function within the web of life. In the human system, diversity is not quite different in philosophy. Diversity creates and stipulates a dynamic environment for economic growth and healthy community because every element and component within its system work together based on various skills, experiences, knowledge, and expertise (Fig. 6.1).

Connectivity

All system is characterized by pattern, structure, and process (Ikerd 2005, 83) that works in terms of interrelationship. In nature, every element and component is connected to each other to build and establish a regenerative and renewable ecosystem (Hester 2006, 48–60). Sustainable system is characterized by a structure that puts elements, uses, positions, and forms in a functional relationship. Interrelationship is another word for relation and communication. It is not only practical but also existentially

Fig. 6.1 Diversity of uses, forms, modes of transportation, populations in Spadina Chinatown Toronto, ON, Canada (Photo by Author)

interdependent for collaboration and habitation. Accordingly, elements in the system are not only related to each other for production, but they also that for distribution and regeneration. In the urban context, interrelationship builds a nexus and network of uses, domains, activities, transits, movements and green areas for habitations, productions, and services (Jerke et al. 2008, 142–77) (Fig. 6.2).

Functionality

There is nothing in nature without function. Ecosystems in nature are distinctive within their spatial units. Albeit the units of the ecosystem are porous and open to each other, the spatial units are not an exchange of mass, organism, energy, and information with one another (Lovett et al. 2005, 2). Functionality in nature is shown by unique activities of energy transformations. It is encompassing vitality and continuity. In order to be functional, the elements of the system are necessarily specific and unique in their role, occupation, and position. In doing so, they can function and find their particular slot and spot for production, service, and distribution. As a functional system, nature and urban settlement are identical in dealing with transformations and changes for production, habitation, and regeneration (Fig. 6.3).

Fig. 6.2 Malacca riverfront (Photo by Author)

In the urban context, functionality is represented by particular use, specific form, and unique design. However, functionality in the human system is not simply practical, but also normative in terms of cultural values. This includes functional domains for public and private use as well as the interface between them. Functionality helps people to know how to use and deal with forms and domains. Indeed, design plays an important role in bringing forth functionality in a communicative way. Design helps people to understand the use of implements as well as their particular relation to the whole system. The task of planning and design is not only to provide the appropriate form and place for a human being but also to integrate the form and content as one. In doing so, sustainable principles for land use planning and urban design are to create and define the functional framework of domains and forms that ensure sustainability. In the urban context, functionality is mainly about the economy and society for urbanity. Planning and design are development tools that ensure

Fig. 6.3 Green open space in Taman Suropati, neighborhood park, Menteng, Jakarta, Indonesia (Photo by Author)

this functionality. All these are necessarily within the framework of the sustainable built environment.

Reciprocity

All components and elements in nature work together in reciprocity in terms of interactions. Reciprocity in the ecosystem is another word for mutuality, interdependence, and collaboration in which the recycling of energy and resources takes place indefinitely. All elements or components work together in a sustainable system to maintain low entropy and balance between decay and renewal rate. Because of reciprocity, an organic community sustains the principle of energy flow through the web of nutrients (Smithson and Atkinson 2002, 442). The phenomenon of mutuality gives us the idea of filling out the gaps by which demand and supply come to terms in a congruent way. Moreover, reciprocity underscores uses, forms, jobs, materials, and designs in an interactive relationship. Reciprocity in nature is a natural mechanism for defying greed and domination. In the urban context, the terms of reciprocity are set out as laws and regulations that protect people from exploitation and dominance. Planning and design for sustainable urbanism are to ensure that there are safe, healthy, and attractive places for human interactions.

Perpetuity

All elements and components in nature are in the flow for sustaining the process of energy transformation. The relationship among elements, uses, forms, and occupations in the sustainable system shows a continuous process toward growth and regeneration. A sustainable system in nature does not work only for a cycle, but for generations. Perpetuity in nature is a necessary process of evolution which is characterized by natural selection, succession, and colonization (Choras 2009, 109). It is such in forest ecosystem based on the equilibrium between the standing quantity of increment and allowable cut (Ciancio et al. 1999, 60). For the human system, a perpetual system of habitat is essential because people need a stable and healthy environment for their home. This includes predictable seasons and climates that enable mankind to cultivate the land and its resources as well as to manage their activities for work, live, and play without any threat of natural disasters and casualties.

Recoverability

Any natural disaster in nature is theoretically recoverable by itself. It just needs time for the gradually recovering process. However, the habitat of a particular type will be preserved if the habitat decay rate is appropriately made by habitat renewal in equilibrium (Schmitz 2007, 123). The capacity to recover by itself is an integrated part of the evolution process and adaptation to new condition. This implies that recovery and rehabilitation from a natural disaster are not to achieve the original form and formation, but most likely to maintain and care for the structure and function of the ecosystem that works for new conditions and circumstances. This includes the possibly vanishing some species and the newly emerging ones in terms of evolutionary process, through genetic mutations or other transformations.

For the human system, any natural disaster is actually a wake-up call as well as an alarming sign that the relationship between man and nature is in crucial condition and in need of serious attention and action with an ecological approach. Indeed, most environmental crises are undeniably caused by human actions and activities which are excessively incompatible with natural laws and orders. In other words, human civilizations are in part and at large responsible for the natural disasters. Thinking of sustainable urban design is an attempt to renew our approach to the relationship between human and natural system by planning and design.

Urban Environment as Human Habitation

The above-mentioned phenomena of the sustainable system are necessarily elaborated for the human system. The man-made environment has to establish an artificial web system of energy transformation for human beings and other beings. The purpose of this web is to keep a low entropy of such a system. In other words, urban environment as habitat is not simply an abiotic and biotic system of nature. Rather, it is a socioeconomic environment and political institution with complex and expandable interactions. The dynamic system of the urban environment is characterized by the will and conscience of man. The interplay between these two human capacities brings about the dynamics of urban development toward economic growth without the depletion of natural resources in quality and quantity.

The sustainable built environment is not simply an ecological matter and problem. Rather, it is a comprehensive system of continuous development of civilization; this is based on the management of human will and the alignment of conscience for the harmony between economic growth and natural conservation. In this respect, environment and economy have to work together to find a common ground of sustainability. For the human system, sustainability is also a cultural problem because it is about managing behaviors and lifestyle that are fit and compatible with sustainable principles mentioned earlier. Then, the phenomena of diversity, connectivity, functionality, mutuality, perpetuity, and recoverability are technological instruments of sustainability which are derived from the natural ecosystem as a model for a conceptual framework. However, the problems of sustainability in the human system are, of course, more than those because human settlement in an urban context is dynamically economic and characterized by the necessity for safety, proximity, dexterity, and civility.

Public Safety

Public safety is the most fundamental for socioeconomic sustainability in an urban context (Mega 2005, 259). This involves environmental plans and designs that create and define domains with crime preventive considerations. Accordingly, domains should be designed with visually and conspicuously ownership and control. In doing so, urban blocks, zones places, and spots are designed to defy any residual and uncontrollable areas. Public safety includes care for: accessibility, protection of under construction, pedestrian crossing with the vehicle, plants, vegetation, and lighting in the night. Fears of crime come up from the unused public spaces which are mostly vacant of activities at particular times of the day and evening. The unused streets and open spaces in the evening in the employment districts create fears of crime if they are not provided with sufficient lighting.

In addition, the remoteness or isolation of employment district from adjacent neighborhoods creates areas without informal surveillance. Such areas are an easy target for graffiti vandalism. It is commonly experienced in an urban area where proprietary interest in the public place and social interaction increases, fear of crime and violence comes to diminish. Sustainable land use planning and urban design are to support vibrant, attractive, and pleasant urban life in public spaces. The phenomena of gaining access and committing a crime in urban areas are less likely when the design of the built environment is able to generate and provide the lively urban environment.

Proximity

Urbanism is characterized with the necessity for being in proximity (Castell and Susser 2002, 64). Proximity in the context of settlement is community living that enables people to intensify their interactions, exchanges, and collaborations. Architecturally speaking, proximity is one important characteristic of the urban environment; this nearness is demonstrated with the phenomena of compact and dense built environment. Moreover, proximity is experienced as the state of being that everything for daily needs is mostly at the reach of hand. The phenomenon of urbanism is historically signified with highly concentrated buildings and populations so that dwelling and settlement are efficiently achieved. This architectural form of urban settlement stands out from the natural landscape. The urban environment is visually and eco-systematically not a natural habitat, but a socioeconomic and cultural system of production that is sustained by human laws and orders. The necessity for compact and geometric form shows the human condition that built environment is technically a reasonable form and domain. In doing so, urban settlement depicts human capacity in organizing forms and domains for the efficacy and effectiveness of habitation. Historically speaking, proximity is the physical character and property of the urban settlement; ancient towns and cities had been mostly erected with compact structure since Jericho of 9600 B.C. and Mohenjo-Daro of 2500 B.C. It is intended to intensify exchanges and interactions for socioeconomic growth of its urban society (Fig. 6.4).

Dexterity and Specialization

Unlike in nature, people as elements of the human system need to be trained and educated to function in their world. Dexterity is about skill and adroitness. In the broadest sense, it includes professional specialization in various fields of occupation and profession. Urban dexterity is characterized by certain standards and codes related to specialist fields of skill which are not only to deal with concrete things but also with people and abstract things and concepts. Because of the sustainable learning process, urban people are culturally and socially sophisticated by their use

Fig. 6.4 Proximity and outdoor hangout in Skudai, Johor Bahru, Malaysia 2012 (Photo by Author)

of language, behaviors, and manners. Dexterity in an urban area is represented by various uses and its particular types, kinds, sizes, sorts, and qualities that reflect a variety of occupations and professions. Spatially, this variety includes and is not limited to the spatial units and zones of residential, industrial, center, employment, commercial, recreational, green, and open spaces. The purpose of land use plan for sustainable development is to create and establish diverse and complementary activities that work for economy and environment. All this is for what we call urbanity. It is actually the essence of urban development that brings about purposes of all urban activities as urban intentionality.

Civility and Urbanity

Unlike in nature, transformations and changes are not simply for succession. Rather, they are in search of perfection, beauty, and habitation. All these are comprised of the concept of civilization. Civility is the process and product of urbanism that cultivates urban activities as a culture. Without civility, there is no urbanity. Civility is implemented practically in the urban life-world which is formulated and conceptualized as laws, regulations, codes, manners, and etiquettes. Indeed, civility is not a product of overnight, but it is cultivated from ancient time until today. As process and product, civility shows itself as what we call urbanity. Thus, urbanity enacts as the intentionality toward which all efforts and accomplishments are directed and motivated. It is not simply economic, environmental, and cultural. It comprises most of everything urban. Urbanity as the outcome of civility is established, maintained, and sustained by urban dwellers as civil society. Sustainability urges civility to go further which is not only for unveiling humanism but also for going deeper into the totality of beings as a whole in a responsible manner.

Prospects, Problems, and Constraints of Sustainability

Sustainability in the Age of Global Economy is not without problems in terms of policy in theory and practice. Obviously, the most problematic matter is about our materialistic and consumptive culture and civilization. In regard to environmental sustainability, the today culture has been relying too much on non-regenerative energy and non-recyclable products. Needless to say in the Western world, our automobile-dependent lifestyle and the American dream of suburbanism is probably an acute resistance against the compact community and public transit-oriented urbanism if it is not another kind of conservative ideology of liberty and free market economy. Indeed, from a historical perspective, suburbanism is in part a by-product of dysfunctional and decaying downtowns and urban centers. Undeniably, suburbia is ideologically a realization of the American dream of freestanding property, new frontiers, and unlimited consumption (Sorkin 2009, 165). The problem of urbanism and suburbanism is obviously ideological. Its ideological dichotomy does not bring about a viable solution. The question is in what way and extent sustainability able to reconcile the dichotomy of urban and suburbanism? Even though a successful precedence of reconciliation between urban and suburbanism is barely to find, it does not mean that it is impossible. Sustainability urges such possibility technologically. This includes planning and design principles for density, connectivity, and diversity, as mentioned earlier.

Since energy dependency on fossil resources is inevitably no future, sustainable resources are ideologically imperative. It is an integrated part of the solution for saving our planet and for the improvement of our quality of life. This is not simply a technical matter. Rather, it is ideological and political because sustainability is about collective awareness of 'one-world' reality and therefore everyone is due to do their part responsibly. Functionally speaking, domains and buildings are meaningless and helpless without any connection and communication, the transit system is part of the sustainable approach to well-functioning planning and design. It is inevitably important to note that an environmentally friendly transit system is the backbone structure that enables us to diminish the dichotomy of urban and suburbanism. Imagine if all centers are well connected by a reliable and sustainable mass rapid transit system. Then, transit-oriented development becomes to be one of several viable options for reinvention and rejuvenation of urbanity, regardless of its location whether in the urban area of in the suburban region. Accordingly, urbanity should be open in the context of a regional metropolis. Presumably, a well-organized transit system should have been laid on with various modes and means so that in case one system is in dysfunction we still have some others as options and backups. Then, transit-supportive public policy for the land use plan and urban design is necessarily interregional, beyond municipality boundaries.

Furthermore, sustainable land use planning and urban design are necessarily thought of as a well-integrated system of public transit network and pedestrian-friendly community. Sustainable urbanism needs environmentally friendly public transit because mass transportation encourages casual interactions and energy conservations. This integration is not only environmentally beneficial but also socially and economically functional that brings forth and develops urbanity toward a new direction for the coalitions and collaborations of environment and economy.

The importance of sustainability for land use planning and urban design lies in its primary concern for the physical environment. Economic growth is doubtful to solve socioeconomic problems and environmental sustainability (Erhun 1998, 147). Likewise, economic growth is also doubtful in the future without environmental sustainability. Economy and environment are necessarily seen as complementary partners that need each other to sustain healthy communities. Land use planning and urban design policy should bring economy and environment together. This is achievable by bringing forth compact, diverse, and aesthetically well-designed man-made environment for safe and healthy communities.

The necessity for economic sustainability is to deal with the fact of increasing depletion of non-renewable resources. As non-renewable reserves are depleted, the price will rise, utilization will decline, then renewable resources and systems will be competitive (Ikerd 2005, 61–2). The sense of sustainability follows the logic of economy as the management of scarcity of resources. Since today clean air, water, and land are polluted, sustainability of such resources is in demand. And its sustainable growth and development are indispensable environmental, in regard to the exhaustion of fossil fuels and global warming. Urban planning and urban design have to do something about this with a new approach that redefines and restructures its thinking, policy, theory, and practice.

Without thinking on sustainability and business, as usual, urbanity is at stake in a self-destructive way. Nevertheless, environmental crises are not out of human commissions and provisions. Therefore, it is also an opportunity for mankind to restore and realign their built environment wisely in accordance with ecological principles of the ecosystem. There are some options to deal with the urban environment in a sustainable way in terms of strategy for policy implementation. Even though aesthetics has been regarded as an integrated part of sustainable land use planning and urban design, its problems and contributions to public policy have been paid with less attention and deliberation. Actually, planning and design for sustainability are inescapable to the necessity of public aesthetics. Unattractive urban form and environment are politically unmarketable either for ordinary citizens or for business people.

Architecturally, the aesthetics of the urban environment comprises a form, material, function, and site. In dealing with these components, the design is to integrate them into an urban system that is developed with the capacity for reducing pollutions, recycling materials and energy, and reuse resources as efficient as possible. The sustainable urban form is not only to use the land efficiently and to protect natural resources on site wisely but also beautiful to integrate man-made environment into its local site and surroundings. Even though the compact urban form is not only the way we achieve sustainability (Williams et al. 2000). However, the compact urban form should be seen and considered as part of the environmental solution that protects the excessive use of land and gives more space for biodiversity and open spaces. Needless to say, the importance of compact built form is to optimize the land and its infrastructure for the pedestrian-friendly environment. In other words, the compact urban form is a sustainable way we design the built environment provided that its integration with its natural locality and surroundings plays a crucial role in the preservation and conservation of resources and healthy environment.

Ecologically integrated urban form with the natural site is not simply a beautiful structure with planting and gardening. Rather it is the act to put natural elements—water, plant, open air, rock, and metal into a built environment so that enable people to remind what nature is on a daily basis. Aesthetic experience in terms of sustainability is the pleasure of intimacy with natural and local elements—rocks, soils, plants, vegetation, and other landscape features. Being sustainable for aesthetics is bringing back people to recognize their relation to the nature of locality. In doing so, people are able to trace back elements and components of form and materials to their origin in the local environment. The purpose of this is to develop the knowledge of materials and techniques that enable people to reuse and recycle without hazardous implications for soil, groundwater, noise, and air. Indeed, most organic building materials are aesthetically acceptable and favorable for sustainability. However, industrial products are qualified for sustainable environment provided that their possible reuse and recycle are environmentally friendly with fewer risks and consequences for degradation of soil, water, and clean air.

Theoretically speaking, the aesthetics of sustainability is the way we are grateful for what we receive from the natural resources; water/rain/snow, air, daylight, and rock/soil/earth. Accordingly, form, material, function, structure, and technique are nothing but the representation and expression of human thankfulness to nature in terms of preservation and conservation in a sustainable way. All design we create for sustainability should reflect our awareness of the gratitude to nature in a way of beautifying the environment. In this sense, green building is neither a fashionable form nor stylistic expression, but an attempt to manage the abundant resources we receive in daily life for being able to be reused, recycled, and

Fig. 6.5 Vibrant, ample, continuous, and green pedestrian environment, Orchard Road, Singapore (Photo by Author)

transformed efficiently. This includes reducing and minimizing the side effects of productive activities and other processes.

Nonetheless, sustainable aesthetics is not simply a pleasure of visual experience on environmental quality of living, working, and playing. Rather, it is about the experience of being thankful for the natural resources we use with knowledge and wisdom that we are able to send and return them back with efficiently energy conserved forms, recyclable materials, climatically adaptive structure, and locally adaptive building system. All this is brought forth and constructed not for self-gratification, but simply as the act of gratitude to the nature that gives us their resources, challenges, and opportunities for being on earth generations to generations. Nevertheless, sustainable aesthetics is ideologically ecological that concerns primarily with healthy indoor and outdoor environment (Fig. 6.5).

Implementation

Prior to the making of any policy, plan, and design, a shared vision for sustainable land use plan and urban design is necessarily created and developed with collaborative work involving all stakeholders that will address current and future needs and desires. The purpose is to advance the comfort, attractiveness, social, cultural, and economic vitality of a sustainable land use plan and urban design through cooperative discussions and public participation. It is also to create a forum for open discussions, idea generation, and research relevant examples and to advance good planning practices by involving the community, neighborhood associations, landowners, retailers, development groups, and public and private interests. The formulation and conception of the shared vision will establish a comprehensive framework that encompasses the various policies, regulations, and strategies to create a coherent and realistic vision that fosters the new development and appropriate redevelopment for the following principles and goals:

- Sustainable land use plan and urban design development should establish a long-term strategy for the healthy and manageable environment;
- Sustainable land use plan and urban design should ensure a balance of needs of stakeholders in terms of economy and ecology;
- Sustainable land use plan and urban design should encourage a sustainable community environmentally and economically;
- Sustainable land use plan and urban design should create a pedestrian-oriented environment with less automobile dependency;
- Sustainable land use plan and urban design should promote a transit-oriented community with safe and convenient transit system;
- Sustainable land use plan and urban design should encourage a mixed-use intensification for residential, commercial, and employment activities;
- Sustainable land use plan and urban design should create vibrant main streetscapes and other public places for safe and healthy social interactions.

A long-term strategy should allow for the vision to be achieved incrementally over time. A comprehensive study is required that will review important community features, alternatively built form types, appropriate development standards, heritage resources, suitable land use models, planning policies, transportation plans, and streetscape design. Such a study should also establish and engender a process that considers urban design, land use planning, transportation system, and community uses. This includes extensive collaboration with the various community groups, residential associations, local Business Improvement Area, and other agencies. Moreover, such a study on sustainability is to provide a framework to stimulate appropriate development, encourage neighborhood growth and pride, recognize the character of the community, and enhance the existing main street commercial strips. Characteristically, a long-term strategy should have the following contents:

- A long-term strategy should be designed to incorporate the sustainable land use plan and urban design in stages;
- A long-term strategy should be comprehensive and interdisciplinary in its approach and implementation;
- A long-term strategy should meet the development standards and accommodative for various building types;
- A long-term strategy should stipulate and take into account the local heritage and resources for its planning and execution;
- A long-term strategy should reckon aspects and factors of urban design, land use, transportation, and community uses as an integrated framework for its performance.

In order to take plan into action, principles are necessary to be well formulated. One important strategic plan and design is to restructure and reorganize urban blocks or domains to address centers. Sustainability for urbanity is inevitable to improve and redesign urban blocks as environmentally friendly spatial units. This comprises optimization for biodiversity and area of precipitation. In a sustainable way, urban blocks should integrate natural and man-made environment within an environmental system that conserves the use of energy and optimize natural daylight for lighting and heating system in the winter time.

Moreover, urban form with a block system should allow natural light and air circulation to work well for human well-being. It is useful to implement urban blocks as spatial units of ecosystem function. Indeed, urban block as a spatial unit is not simply an environmental system, but social and economic organization. However, as ecosystem function, the urban block is not simply abiotic environment; artificial and natural component should be intertwined as a cohabitation system. The integration of biotic elements and components is necessary to optimize the process of reduction, recycle, and reuse resources in a sustainable way.

Architecturally speaking, urban blocks should avoid mono superblock buildings which are more dominated by abiotic components and elements as well as by impenetrable structure for the pedestrian easement. In addition, sustainable urbanism is to create healthy citizens with the more pedestrian-friendly outdoor environment so that pedestrian permeability of urban blocks is part of the concept and implementation of sustainability (Fig. 6.6).

Principles of sustainable land use and urban design should include the mandate to support site remediation; reduce energy consumption, green building design; compact built-form development, efficient mixed-use buildings, improved air and water quality, expandable and diversified public transit; protect natural and human-made resources, efficient use of infrastructure; encourage gray-field redevelopment; promote public health and safety; support the local economy for long-term prosperity; and create a walkable and bicycle-friendly community.

Moreover, there are some considerations for pedestrian design and construction as follows:

- Make the most of the caption area for precipitation with permeable paving materials or natural ground covers;
- Keep and preserve existing significant trees and vegetation and incorporate them into the site design of new development;
- Select and plant new trees and vegetation with less water ingesting;
- Provide landscape design to break the impermeable area into smaller landscaping areas that minimalize stormwater discharge into the storm drain system;

Fig. 6.6 Pedestrian-friendly environment in Orchard Road, Singapore (Photo by Author)

- New development should meet and be in compliance with LEED or Green Globes Certification for environmental responsibility and energy efficiency;
- Reduce reliance on non-renewable resources in building construction and use suitable recycle building materials;
- Make most of the use of daylight and low emission finishes and green roof in building design;
- Minimize energy consumption indoor and outdoor building system with the new method and technology available in the market;
- Improve water treatment system with appropriate water management and efficient plumbing system;
- Where possible reduce, recycle, and reuse materials from the devastation of existing buildings for new construction;
- Protect and hold existing trees, vegetation, natural slopes, and native soils and integrate all these into a system of landscape plan;

- Link all green open space in a continuous system with pedestrian or bike cycle network;
- Make most of shade and stormwater benefits by distributing landscaping throughout the area for pleasant pedestrian and cycling routes;
- Combine soft landscaped areas to enhance tree and plant material growing conditions.

Buildings on traditional main streets are typically designed with active retail spaces on the street level. The retail areas are combined with either residential or office uses on the upper floors. Sustainable land use plan and urban design are necessarily provided with the hybrid of economic and office or residential activities. In order to maintain its diverse service times, sustainable urbanism is generally composed of retail buildings with direct accesses from the sidewalk of the street or square. To create a vibrant around the clock environment, multi-use buildings should be encouraged along main streets. Unlike commercial superblock, the sustainable livability of urbanism needs independent small block buildings with direct accesses from the public realm. Small blocks create a human scale and vibrant ambiance. In the case of special occasions, the street space can work as a pedestrian environment for intimate social interaction of people (Fig. 6.7).

Multi-use buildings add vitality to the street by integrating live, work, and shopping uses; encourage less automobile use by reducing travel needs; create an efficient, compact, and sustainable built form; supports public transit usage by encouraging appropriate intensification; can reduce parking needs by shared arrangements; and can respond in a more flexible and adaptable manner to meet future economic and market needs. Some considerations for vibrant urbanity include and not limited to:

- The uses of buildings along the main street should be functionally complementary of the shop, live, and work facilities with a single system of development;
- Small retail commercial units should encompass a compatible mix of uses that promote an efficient pedestrian network;
- The use of the upper floor should be different from the use on a ground level. Nevertheless, the second level may be designed to have the same

Fig. 6.7 Street space and Night Market in Malacca 2012, Malaysia (Photo by Author)

use as the ground level as long as the building comprises, at least, one floor above the second floor;

- Ground level should be given priority for commercial uses with a direct entry from sidewalks;
- Reassure a better balance of land uses for a living, work, and shop;
- Endorse timely provision of public infrastructure needed to encourage private investment to preserve and revitalize the neighborhoods;
- Encourage efforts to provide needed services such as child care, youth programs, and elder care in main street activities so that people have time to join in community events, planning, and decision making;
- The eventual goal of sustainable land use plan and urban design is to make, establish, and sustain vibrant, pedestrian-friendly, and active main streets. This could be achieved with optimization of resources on both sectors—private and public. Revitalization of main streets means

also to create a conducive environment for the collaboration of both parties mentioned above.

By advancing the other goals such as community development, public places will become the heart of the community and engage the larger neighborhood to contribute and benefit from the revitalized character of locality a vital part of the community that is attractive, safe, and accessible; contribute positive social, economic, and environmental benefits; and improve public health, well-being, and quality of life. The other enhancement for vibrant street activities includes and not limited to:

- Provide active frontages and legible entrances with various retailers or commercial uses at street level;
- Find more retailers with possible activities spilling out from inside to the sidewalks;
- New expansion in a sustainable land use plan and urban design should improve the adjacent boulevards and sidewalks by integrating pedestrian lighting, street trees, decorative paving, and street furniture where applicable;
- Optimize the area for soft landscaping in both the public and private realm;
- Where applicable, provide continuous street tree planting with an interval of 6–10 m or trees in clusters;
- Plant high branching deciduous trees to create a canopy and provide shade in summer time;
- Use a variety of trees and ornamental shrubs on private property adjacent to the street for year-round attention;
- Discourage strip centers or single-story buildings with deep setbacks, excessive curb cuts, frontage parking lots, and drive-through services.

The pedestrian-friendly environment is an important part of the sustainable land use plan and urban design because it contributes to healthy citizens and urban environment. The pedestrian-friendly environment could decrease the use of space for hard landscaping with impenetrable materials. In doing so, precipitation of land is made possible to increase in its captive area. However, the notion of the pedestrian-friendly environment should include functional, safe, and attractive design of area for pedestrian movements and amenities. The pedestrian-friendly environment in the public realm should be a well-connected network of pedestrian routes and open areas. Some other considerations are useful such as:

- Access to main streets should be limited to existing signalized intersections. Restricted right in/right out accesses along the arterial road are not permitted. These arterial access/intersection points will be assessed in more detail at the development review stage;
- Additional access points along main streets should not be encouraged;
- Pedestrian space along the main street should be designed as a prominent corridor with the high-quality performance of form and materials, supportively promoting active use by residents and visitors;
- Pedestrian space of the main street should be rich with amenities such as street furniture, public arts, banners, lightings, plazas, and historical features that should create and establish a sense of place;
- Sidewalks along the main streets should be designed in appropriate scale to the adjoining land uses. Ample sidewalks should be provided at the commercial areas where pedestrian activity is high;
- The safe pedestrian environment requires unobstructed and free of hazards pedestrian movements along and across the street;
- Nevertheless, sufficient lighting and buffers from fast-moving traffic should be included in the pedestrian-friendly environment.

Sustainable land use plan and urban design are never effective without the support of reliable and convenient public transit system. The reason is simple because public transit reduces the use of private cars in terms of the need for parking spaces and emissions of CO and CO_2. The notion of transit-supportive design is a scheme and project that provides a development framework that supports the increase in public transit ridership. To achieve this development should support densities and a range of complementary uses—live and work. Along main streets where pedestrian activity is high, transit facilities should be convenient to use and discourage private automobile dependency. In order to generate high pedestrian activity, higher residential density along the main streets should be created and established with the following consideration:

- Transit facilities should be located within a five-minute walking distance—approximately 350 m—of most residential and commercial uses;
- Discourage uses and developments with automobile dependants such as drive-through and gas stations;
- Minimize automobile accesses to facilities along the main street. Land consolidation is encouraged to facilitate developable land parcels with consolidated rear parking lots;
- Promote bike cycle routes and trails which are well connected to main street activities;
- Provide secure and convenient bike racks and parking space at the areas with high pedestrian activity;
- On sites adjacent or close to transit shelter, the provision of a safe and pleasant walkway should be provided that connects and crosses individual sites within and between blocks, zones, and points of interest;
- Building entries should be oriented and directed toward transit facilities for convenient access by public transit passengers.

Concluding Notes

Sustainability for land use planning and urban design is neither a new trend nor style, but simply a necessity to deal with urbanism with respect to the healthy environment. Theoretically, sustainable urban planning and design can be seen as self-critical recourse of urbanism; it is in dealing with the environmental crises and depletion of non-renewable resources. In achieving sustainability, land use planning and urban design as public policy are not inclusive if they are not well coordinated and integrated with an environmentally friendly public transit system and sustainable urban economy. Sustainable land use plan and urban design should bring economy and environment in dialectical partnerships and collaborations that find their common ground in compact, diverse, well-connected, and aesthetically well-designed built environment for work, live, and play as a system. Nevertheless, the sense of urbanity is barely achievable only through a public policy for environmentally friendly urban form. Urbanity urges an interdisciplinary approach and practice of public policy that brings forth and develops vibrant and healthy communities. The task of urban planning and design for sustainability is to integrate activities, forms, and domains as environmentally friendly and transit-supportive places and communities.

References

Bairoch, Paul. 1991. *Cities and Economic Development: From the Dawn of History to the Present.* Translated by Christopher Braider. Chicago, IL: University of Chicago Press.

Beatley, Timothy, and Peter Newman. 2009. *Green Urbanism Down Under: Learning from Sustainable Communities in Australia.* Washington, DC: Island Press.

Castell, Mario, and Ida Susser. 2002. *The Castells Reader on Cities and Social Theory.* New York: Wiley-Blackwell.

Choras, Daniel D. 2009. *Environmental Science.* Sudbury, MA: Jones & Bartlett Publishers.

Ciancio, Orazio, Piermaria Corona, Francesco Iovino, Giuliano Menguzzato, and Roberto Scotti. 1999. "Forest Management on Natural Basis: Fundamentals and Case Studies in Piermaria Corono." In *Contested Issue of Ecosystem Management,* 121–56. Pennsylvania, PA: Haworth Press.

Erhun, Kula. 1998. *History of Environmental Economic Thought.* London: Routledge.

Farr, Douglas. 2008. *Sustainable Urbanism: Urban Design with Nature.* Hoboken, NJ: Wiley.

Friedman, Avi. 2006. *Sustainable Residential Development: Planning and Design for Green Neighborhoods.* New York: McGraw-Hill.

Gouldson, Andrew, and Peter Roberts. 2000. *Integrating Environment and Economy: Strategies for Local and Regional Government.* London and New York: Routledge.

Hester, Randolph T. 2006. *Design for Ecological Democracy.* Cambridge, MA: MIT Press.

Ikerd, John E. 2005. *Sustainable Capitalism: A Matter of Common Sense.* Sterling, VA: Kumarian Press.

Jacobs, Jane. 1961. *The Death and Life of the Great American Cities*. New York: Random House.

Jerke, Dennis, Douglas R. Porter, and Terry J. Lassar. 2008. *Urban Design and the Bottom Line: Optimizing the Return on Perception*. Washington, DC: Urban Land Institute.

Lovett, Gary, Clive G. Jones, Monica G. Turner, and Kathleen C. Weathers. 2005. *Ecosystem Function in Heterogeneous Landscapes*. Berlin: Birkhäuser.

Lynch, Kevin. 1981. *A Theory of Good City Form*. Cambridge: MIT Press.

McLennan, Jason F. 2004. *The Philosophy of Sustainable Design*. Bainbridge Island, WA: Ecotone.

Mega, Voula. 2005. *Sustainable Development, Energy and the City: A Civilization of Visions and Actions*. New York and Berlin: Springer.

Newman, Peter. 1996. "Reducing Automobile Dependence." *Environment and Urbanization* 8 (1): 67–72.

Riddle, Robert. 2004. *Sustainable Urban Planning*. London: Routledge.

Schmitz, Oswald J. 2007. *Ecology and Ecosystem Conservation*. Washington, DC: Island Press.

Shachak, Moshe, James R. Gosz, Stewart T. A. Pickett, and Avi Perevolotsky. 2005. *Biodiversity in Drylands*. New York: Oxford University Press.

Smithson, Peter, and Kenneth Atkinson. 2002. *Fundamentals of Physical Environment*. London: Routledge.

Sorkin, Michael. 2009. "The End of Urban Design." In *Urban Design*, edited by A. Krieger and S. Saunders. Minneapolis: University of Minnesota press.

Steele, James. 1997. *Sustainable Architecture: Principles, Paradigms, and Case Studies*. New York: McGraw-Hill.

Thomas, Randall. 2003. *Sustainable Urban Design: An Environmental Approach*. London: Taylor & Francis.

Williams, Katie, Michael Jencks, and Elizabeth Burton. 2000. *Achieving Sustainable Urban Form*. London: Taylor & Francis.

7

Urbanism and the Global Age

Urbanism in the Globally Exposed World

This study is to deal with the issues, constraints, and opportunities of global urbanism in North America that matter for the sustainability of the sense of home. The point of departure for this study is the concept of urban settlement as a home for everyone who works, lives, and plays in the context of socioculturally, economically, and politically organized systems in reference to certain sites on regular basis, home as the life-world. This study understands the life-world as the system of intentional happenings that involves people, domains, and things for specific purposes on regular basis. The focus of the study is to investigate and examine issues, constraints, challenges, and opportunities of the urban life-world in the age of virtually connected society. The purpose of this study is to explore considerable points and findings for drafting urban development and governance policies at the federal, state/provincial, and municipal level in the context of global urbanism. The rethinking of key concepts of global urbanism belongs to the main objective of this study that enables us to have an overview and options concerning the structures and function of agencies within the system of global urbanism.

© The Author(s) 2020
B. Wiryomartono, *Livability and Sustainability of Urbanism*,
https://doi.org/10.1007/978-981-13-8972-6_7

The structures and functions of governance include the dynamic power relations and control mechanisms that work and manage to deal with the daily life-world of urbanism. Methodologically, this study is to deal with the presence of global urbanism in relation to the built environment within the conceptual framework of settlement; to what extent is the concept still valid for the sense of home today or is this concept necessarily rectified due to the contemporary world conditions? The key concepts of locally established urbanism, which are related to the urban phenomena, such as urbanism, urbanization, society, community, citizen, civility, polity, market, and the public have been considerably falling into oblivion if they are not taken into granting in most theoretical explorations.

In this study, the starting point is the concept of urbanism that is a systematic construction of efforts, thoughts, policies, and actions for establishing, developing, maintaining, and sustaining human settlement as a home for everybody who works, live, and play in its spatiotemporal territory. The sense of home entails safe-healthy-attractive and resilient places for residential, sociocultural, economic and political activities. For this study, the question of home is the guiding path of investigation and examination. The path will lead this study to confront the things, structures, functions, agencies, and processes of urban phenomenon in the age of technological information. These confrontations are in the search for understanding and dismantling the issues, constraints, challenges, and opportunities we experience in the life-world of the virtually connected society. To what extent has this virtual reality transformed and expanded the meanings of our global cities? However, it is not the objective of this study to examine and review the available theories on global urbanism in details which are mostly not from the same perspective with this study; the purpose of this study is limited by its specific area of concern, namely the question of a home in the Age of Global Information.

In recent decades, global cities have been extensively studied from various aspects and interests by a number of scholars from various backgrounds and disciplines. A number of scholars have made the concept of the global city more understandable. Our formative and preparatory understanding on global city and urbanism has been founded by Levebre (1991), Hall (1998), Harvey (1998), Short (2014), and Scott and Storper (2015); their contribution to theories of urbanism is invaluable and fundamental to understanding the complexity of urbanizations, urban forms, structures, functions, relations, and domains of the city.

The necessity for in-depth discourse on theories on global urbanism has been well formulated by Roy and Robinson (2015) as a momentum to rethink the Euro-American legacy of urban studies. All this is considered as the necessity in the interest of epistemological inclusivity, and not simply an opportunity in response to the well-documented studies on global city in the Northern hemisphere, which have been presented by Robertson and White (2003), Brenner and Keil (2006), Krause and Petro (2003), and Amen et al. (2006); their studies inform us with global features, traits, constellations, positions, forces, conditions, and problems concerning our urbanism today in the global north.

In North America, global urbanism has been historically identified with the sociopolitically constructed system of settlement that has been economically working around its objective reality as the materialization of the capitalist mode of consumption and production. Roy and AlSayad (2004), Gugler (2004), Dawson and Edwards (2004), Parnell and Oldfield (2014), and Miraftab and Kudva (2015) have enriched and extended our understanding on world cities in the Global South that unfold informality, poverty, sociocultural complexity, and inequality as unique themes and issues of urbanism. Robertson and White's compilation (2003) has enriched and heightened our understanding of the global city with the comprehensive coverage of thoughts on urbanization and globalization from a sociological framework and perspective. Saskia Sassen (1991/2013) has unfolded the emergence and development of global cities from their significant role and function in the world economic hegemony through their network of markets while Castells (1996) has provided us with the characteristics of the global city today as a network society that works in the global space of flows and connectivity. The fact of globally connected society through information technology has enabled us to build a global awareness of the wholeness of our world concerning the various issues, constraints, and opportunities of urbanism, such as climate changes, human rights, and social justice.

Even though the economic and financial system within the capitalist mode of production plays an important role in global urbanism, there are other essential structures and components that have been shaping, developing, and exploring the manifestations and features of the global city from various positions of humanities and social sciences (Abrahamson 2004; Acuto 2013; Lechner and Boli 2012; King 2015; Xiangming and Kanna 2012; Bishop et al. 2003; Birch and Wachter 2011; Calthorpe 2010; Short 2014; Scott and Storper 2015). The contribution of each theoretical approach and treatment toward our understanding of global urbanism is significant from its diversity, profundity, and acuity.

Most available compilations have surveyed various aspects of global cities within the framework of urbanization and in the interest of explorative studies on what global city is. Nevertheless, from the contributors, there is hard to find any line of thought and theory that deals with the relationship between global urbanism and the human condition in the Age of Global Information. While the available theories have utilized extensively the concepts related to the phenomenon of urbanization, such as universalism, particularism, modernism, postmodernism, capitalism, liberalism, and neoliberalism, contemporary scholars have paid less attention to work on the philosophical profundity of the key concepts related to phenomenon such as urbanism, urbanity, civility, and urbanization. Since the streams and points of view of the studies are widely open and explorative concerning the phenomenon of urbanization from a political economic perspective, it is not easy to have a solid and theoretically consolidated understanding of what is global urbanism in the interest of human conditions that involves works, labors, and interactions (Arendt 1958/2013).

In most cases, the theories of global urbanism have been understood within the framework of urbanization as a conceptually fixed construct in terms of the object of the thinking agency; in this study, urbanization is understood as the process and production of a sociopolitically and economically organized human settlement that involves permanently concentrated populations for their works, lives, and recreations. Since urbanism is commonly a locally urbanized system of settlement, confronting local and global urbanism is considered necessary. In doing so, urbanism has essentially experienced in the daily life-world as urbanity that has something to do with the manners, ways, forms, styles, and fashions of human works, labors, and interactions toward a highly cultivated system in terms of civilization.

While most scholars were occupied with the theoretical exploration concerning the impacts and consequences of global capitalism, the sense of home in the globally constructed reality of urbanism remains in need and deserves for further exploration. This is not simply the objectification of explorative thoughts about global cities. This is the ontological necessity of the global life-world that is to exist, build, develop, and sustain the socially constructed urbanity in the age of worldwide information. As a conceptual construction, global urbanity is the essential grains of urbanism that needs and deserves new theoretical elaboration and elucidation in dealing with the specific sense of purpose. This study argues that the theories on global urbanism are necessary to deal and to confront with the sense of home as one of the most important purposes of urbanism. The necessity for the sense of home is considered the existential condition of humanity that enables one to know what, where, and when he/she has to settle things down in the continuously connecting and moving worlds. A theory of global urbanism concerning the question of home enables us to be aware of global issues, constraints, and opportunities concerning our work live and play today and tomorrow.

Urbanism and the Sense of Home

The sense of home in urbanism is experienced as the feeling of being settled with familiar persons and things that are within the reach of hand. The sense of home is the ground of our being (Kennedy 2014, 15). This is the sense that the life-world is safe, healthy, and convenient to work, live, and play. The necessity for home has been philosophically undertaken by Martin Heidegger in his *Building Dwelling Thinking* (Heidegger 1977, 323–339); building a house, village, town, and city is nothing but letting dwell in the search for home. Since for Heidegger the state of being homeless is ontological, therefore, human beings have to learn to dwell within the fourfold: earth/matter-sky/space and divine/eternity-mortal/finitude. This is not simply to have a house. Rather, the sense of learning to dwell is to build an intimate relationship and proximity with locations and things in terms of habitation so that one feels being at peace and harmony with his/her life-world. In other words, the building is bringing forth the four-fold elements and dimensions into a thing as the site of habitation.

Hannah Arendt characterizes the home as a private sphere in the context of the man-made realm where human work, labor, and action take place as a whole we call it the world (Arendt 1958/2013); home in this sense is the private life-world where intimate and proximate relationship of people, domains, and things is established, developed, and sustained with a socially ordering system and domesticity. Since humankind is political and economic beings that necessarily work, live and grow together in peace and harmony with a permanently organized system throughout history, they establish, develop, and grow their notion of the socially ordering system and technologically organized production of the household life-world. The materialization of authority and domesticity in the larger scale of dwelling is represented by the agency of governance and the institution of the market that construct the ontological structures of urban settlement in terms of town and city.

Despite changes and transformations in urbanization, the necessity for the sense of home is an eternal condition of humanity. There is nothing arresting question concerning the global urbanism than the question of a home in the globally exposed connectivity. What is the experience of real-time connectivity that overrules geopolitically demarcated places? So, what is all this for the sense of home? The big cities of North America are such places with a long-standing tradition of the multicultural melting pot. On daily basis, intercultural interactions have been part of urbanism in these cities. However, how this connectivity reaches a far-reaching space-time reality for cultural understanding is one of the most important tasks of urbanism in the Global Age. This issue is not coming from a thin air, as signaled by Samuel P. Huntington's *The Clash of Civilization?* This conflict does not come from primarily ideological or principally economic but predominantly cultural between the West and the rest of the world (Huntington 1993).

Big cities on the globe as the melting pots of various cultures are challenged to deal with harmony and growth in diversity. As a nation-state body, the role and function of the municipality have been challenged by international free trade and democratic internet accessibility that the domestic and territorial authority for politics and economy could have been gradually undermined. Local economic sovereignty and exclusive business relations become unlikely to survive against the transnational capitalist groups and other global communities. The research question is to what extent the municipality is aware of these challenges and sets a strategic political, sociocultural, and economic agenda in their plans and programs in dealing with the movement of global urbanism. Furthermore, this study is to investigate and explores issues, constraints, and opportunities of global urbanism in the context of the municipality in terms of public policies and action plans.

The Genesis of Global Consciousness

What is global urbanism? The question is uncanny in many ways. Some will think it is a question of definition, some are not feeling comfortable and ignore it. Some will take it seriously and ready arguing from various perspectives. What is really matter is that the question becomes indispensable concerning the today phenomenon of cyberspace interconnectivity that brings about new challenges for the concepts of space-time, culture, authority, capital, freedom, and civility; this connectivity challenges the concepts of boundary and demarcation from their association with locality and history.

There is nothing more striking for the concept of global urbanism than the phenomenon of worldwide connectivity. Economically speaking, this connectivity is represented by the presence and network of globe-wide franchised industry and worldwide online connectivity of businesses; these are culturally constructing and shaping our consciousness of what the global urbanism is. In a matter of fact, there is no any capital city on earth today free from global food and technological information accessibility. What is global urbanism is the question concerning societies living, working, and enjoying the sense of standardized taste for foods and drinks and unlimited connectivity for information everywhere on the globe toward borderless inclusivity?

The global consciousness among people on the globe grows with the help of internet connectivity. The 2016 research of the Pollsters GlobeScan as quoted by BBC London (Grimley and BBC 2016) shows that 51% of 20,000 people questioned in 18 countries considered themselves as global citizens rather than as national residents; the 15 year tracking study began in 2001 shows that Nigeria (73% up 13 points), China (71% up 14 points), Peru (70% up 27 points), and India (67% up 13 points) stood out in the highest numbers, while Germany scored with 30% of the respondents who saw themselves as global citizens (BBC World Service 2016).

Global Urbanism and Civility

There is nothing more cognitively striking for urbanism than the concept of civility that recalls and associate people with the forms, manners, features, properties, characteristics, and uniqueness of urban settlement and its society. The concept pertains to the specific world that takes place in an architecturally built environment. Since urbanism is historically a sociopolitically constructed reality, its manifestation has been shaped, established, developed, and sustained for human settlement; civility is associated with society in a sociopolitically organized system. The concept of urbanism is indispensable from the phenomena of the civil world. Under the notion of civility, this study understands it as a concept, which is associated with the socioculturally cultivated behaviors that are practiced, codified, and developed by people living in a place with highly concentrated populations. Civility is considered as the essential grain of towns and cities, in contrast to villages or territorial vicinities. Civility has been historically developed with written codes and laws for social orders, currency, and tax for trades and businesses, and for governance, authority, and administration.

The global society today is the new life-world that is technologically constructed with the globally connected information system of internet network. This connectivity is constrained technologically to embrace people to build a society and public sphere on a spontaneous basis like casual interaction in local urbanism. Video conferences are not limited by its number of participants, but aura and vibration are absent in such meetings. In terms of urbanism, global urban society needs to establish the common ground of their interactions toward civility. How can this sense of purpose be established and developed in the cyber world platform? This is not an impossible thing, but it entails a collective awareness and movement of the urban societies on the globe that they need to set up a common ground for civility in the Age of Global Information. One of the foundations of civility has been laid down by the United Nations with the charters of human rights. In other words, the technological connectivity of societies on the globe does not have the sense of purpose without having global ethics in their virtual life-world.

Global Urbanism, Proximity and Space-Time

This part is to discuss and rethink the concept of space-time in the Age of Global Information technology. The question concerning global urbanism leads us to the question of space and time. The internet connectivity undermines the physical boundary of space and time. This part is to investigate and examine the concepts related to places where people used to work, live, and play in physical geographical setting and environment. How does global urbanism work and take place in the cyberspace-time that matters for the people's lifestyle and way of life in this new paradigm?

Proximity in the Age of Global Information is experienced as the audio-visual nearness based on internet connectivity. This nearness brings people from different locations in interactive communication. One missing part of the experience of cyber connectivity—but very essential for the sense of reality—is the absence of physical contacts, such as greeting with hand-shaking. The sense of closeness or intimacy with physical contact plays an important role in urban society that is experienced as public aura, vibrancy, and animation. The physical crowd in public realm provides us with the sense of liveability. The real livability is distinguished from the virtual one by the tactile experience so that all senses are able to experience the happenings of the crowd with a multitude of senses (sight, hearing, smell, touch, and taste).

The significance of urbanism is demonstrated by the crowd in the public realm. Global urbanism is characterized by the multicultural crowd. These characteristics are demonstrated by the variety of attributes, performances, signs, and symbols from all people over the globe. This diversity is poten-tially growing and enriching the urban society with new cultural products. All this is made possible by the democratic system of the decision-making process with care for the minorities. In other words, global society in an urban context is necessarily considerate for the voice of minorities. This care is considered as the necessity for maintaining the richness of resources, from which urban society can learn for its future development.

In the Age of Global Information, multicultural society becomes socio-culturally constructed reality; people from different places are pulled together to a city because of various reasons. These migrations are commonly for the reason of employment, a refuge from conflict areas, and other personal reasons. Consequently, the global city becomes the locus of the globe where people from various places around the globe are able to grow together. This process is, of course, not without challenges, especially for urban societies with relatively established history, such as those in European countries. To a certain extent, frictions and clashes among people because of their sociocultural backgrounds are not easily resolved with normative procedures of law and order. The alienation and marginalization of a group of people from the decision-making process of urbanism are necessary to be avoided; both are potential to build up sociopolitical extremism and radicalism with the use of violence and terror.

Global Urbanism and Civil Liberty

This phenomenon is not effective in terms of transregional and international communication, but this has become the condition of global business that the resources: Capitals, labors, and commodities are able to move fluidly from and to anywhere in the name of freedom. The experience of freedom is made possible with the exercise of the right of all populations to the city (Levebre 1991), which is based on the supremacy of law and social order in a democratic society. Territorially organized authorities such as nation-states, regional authorities, and municipalities have to face the necessity for a redefinition of individual liberty and privacy in the context of the democratic society of the global city.

The civil world is the globally connected reality of urbanism that is sociopolitically constructed with civic systems consisting of international conventions and trade agreements. The global society needs to establish the common ground for the right to the global city in the context of civil society. David Harvey points out the right of the city as the right of individual access to the resources that the city embodies; this includes the freedom to make and remake ourselves and our cities that is yet most neglected in our human rights (Harvey 2008, 1–2). The right to the city is still a challenging issue either for local or global urbanism. This is not simply about the claim for individual civil liberty. Rather, it is the necessity for the sense of home that is free from fears and threats that include social alienations, diseases, ideological fanaticism, discriminations, and poverties.

The crucial issue concerning global urbanism is challenged by the influx of worldwide migrants into the cities. The challenge is obviously to reveal the concept of citizenship. Even though most Global North countries are demographically declining, immigration is not always the best solution because citizenship is commonly not simply a normative process. Sociocultural adaptation is a most challenging process for the integration of immigrants in the host countries' citizenship. Historically speaking, civic rights in England emerged in the seventeenth century concerning the struggle for individual liberty with respect to the freedom of conscience, worship, speech and the right for making contract and ownership of property; the very idea of citizenship is the equality of all citizens in the eyes of law (Delanty 2000, 15). This concept of social citizenship is not only about equality of rights. For the newcomers of global cities in the North Hemisphere, the common notion of citizenship is associated with the identity of nation-state membership and sociopolitical recognition in the host countries that is bound up with the rights and responsibilities of sociopolitical and economic participation. This is, of course, not a transnational identity and membership that enable people to move everywhere without immigration restriction. The issues, constraints, and opportunities for global citizenship are not about the identity and membership of inhabitant in a community of nation-states. Rather, global citizenship is necessarily regarded as an integrated part of the development of worldwide civil society. In this point, Anthony Gidden points out that globalization is not the same with internationalization; it is not just about closer ties between nations, but concerns processes, such as the emergence of global society, that cuts across the borders of nations (Gidden 1998, 15).

Global Urbanism, Market, and Capital

Today capitalist mode of urbanization has been playing a significant role in the formation, development, and sustainability of urbanism in most everywhere. The global capitalism that has been shaping and dominating the cities on the globe has been a long historical development of European mercantilist imperialism and colonialism since the sixteenth century or earlier that established the global networks of trade and commerce until today. Urbanization in industrial countries has been involving and being resulted from the dynamics of intertwining power between capital and labor. Democratic societies in the Global North have been managing to minimize the consequences of capitalist hegemony on urbanization. The danger of this hegemony is the accumulation of capital power that brings about the welfare state and social justice in peril. So that our concept on capital is necessary to be redefined in the broadest sense of the word as cultural (wealth of knowledge), social (wealth of networks), and symbolic power that includes all of its resourceful potentiality for interest-oriented actions (Bourdieau and Thompson 1991; Bourdieu 1986). Accordingly, the notion of capital should be not understood merely as a financial power and economic resource. Rather, capital should include political, social, and cultural power and resources that are able to converse as symbolic power. Since the various forms of capital: social, cultural, political, and economical are intertwined together within the practices of governing (Olson 2008).

Theoretically speaking, the conversion of all forms of the capital is subject to governmentality that has been historically playing a significant role in the political economy of the urbanization in the Global North. Since the political economy from the global perspective has been supplemented into governmentality, the territorially established governments, from nation-states to municipalities, have been challenged to negotiate and compromise with the global market forces. The global capitalist free market economy leads cities in the northern hemisphere to be entrepreneurial urban centers. Such cities are characterized by competitive and creative talents with a conducive environment for technological innovations and investments, job specializations, and strong private sector initiatives (Florida 2008).

In the context of the Global South, global capitalism and liberal democracy are conceived by the mainstream approach to urbanism as ordinary, pervasive norms and capable of overcoming the poverty, inequality, and injustice perceived as so pervasive across the Global South (Sheppard et al. 2013, 3). In a matter of fact, capitalism and democracy are not always able to defend free market economy from financial crises and political instability. The international community needs to work together for politically integrating national economies from every part of the globe into a global economy of expanding trade and financial flows.

Urbanism today needs to deal with the reality of the global capitalist free market economy. The worldwide power, networks, and flows of capitalism are the challenging issues, constraints, and opportunities for urbanism today, especially on the relationship between the governmentality and the market. The impacts of global capitalist hegemony are experienced as living and working in uncertainty and in predictability. This is actually the crisis of the sense of home. Everybody everywhere is in the economically shaking life-world. To what extent ideological contestations between antagonistic approaches to political economy bring about urbanism toward stable growth and healthy development?

As an ancient institution, the market has been a platform for trading where buyers and sellers interact and transact directly without a middleman. In its traditional form, the marketplace is a core of human settlement with the dense social network. The market is an economic institution that works to create an orderly structured system and reduce uncertainty in exchange (North 1991). A market is an economic event of supply and demand interaction that takes place for a fair bargain. Open competition for better price and service characterizes the nature of the traditional free market. The more complex is the society, the more diverse is its market system. The market grows from time to time in its scope and size of commodities and services.

Then, the concept of the market has been evolving from a geographical fixed place to anywhere that the intercourse of supply and demand takes place. In the broadest sense of the word, the market is a platform for exchange between supply and demand that involves interactions and transactions with more viable and secure. At this point, market needs to be integrated into a larger political institution that protects and secures its event. In order to secure the values of their transactions, sellers and buyers need legally binding process and documentation. For this purpose, lawyers, banks, and brokers become an integrated part of the economic market system; they play their role as intermediaries of interactions and transactions between sellers and buyers.

In the Age of Global Information, the market is characterized by a digital system of interactions and transactions. The transaction record or ledger is called blockchain; it is a part of the transparent and continuously growing list of record of data. The data are stored with a digital code in a transparent way and shared and protected from deletion, interfering, and revision in databases. Typically speaking, each block contains an encrypted hash of the previous block. All blocks are linked together with peer-to-peer computer servers that build a platform-wide and continuously synched ledger. All information in this blockchain is secured by an algorithm with mathematical code and stored, replicated, and continuously updated in every computer within the network so that it is transparent but cannot be forged and to tamper without anyone knowing. Technologically speaking, blockchain is motivated by the idea of efficiently decentralized and democratized database and general ledger by removing unnecessary intermediaries.

Blockchain allows people to transact and interact directly so that every contract, every task, every process, and every payment would have a digital record and signature. This record could be recognized, authenticated, stored, and distributed across several computers. In the future, mediators like lawyers, brokers, and bankers might no longer be required. The blockchain technology transforms the way we interact and transact with more flexible and convenient institution. With its peer-to-peer network, blockchain becomes a registry of the database for assets and transactions. As a new economic institution, blockchain brings about a new platform of economic activities with the more predictable outcome. Until recently, the use of blockchain technology has been playing an important role in replacing paper-based transaction processing. Some country's financial institutions have been testing a digital currency for inter-bank transfers. In a matter of fact, blockchain technology is able to substitute the expensive and inefficient bookkeeping and payment networks of the economic industry. In the Age of Global Information, the complexity and diversity of the market are enormous. The novelty of internet technology enables the blockchain platform growing its application in various fields, such as the voting system of public election, medical records of health care, and vehicle registration.

Global Urbanism, Media, and Culture

Healthy, safe, and attractive urbanism is socially dependent on its society as a whole. This part of the study is to dismantle and unveil the ontological structures and functions of urbanism that gather people together as citizens, urban societies, communities, and interest groups and professional associations that establish, develop, maintain, and sustain the livability and growth of human habitation. The potentialities of multicultural resources and newly emerging cultural movements bring about the enrichment and enhancement of sociocultural quality of urban life based on the network of global urbanism and cooperation. The diversity of multicultural populations and their environments enables them to develop new ideas and techniques as well as knowledge and skills for a strong community; all this is provided that intensive interactions and learning process are well accommodated and appropriately facilitated. Creating a vibrant public realm for multicultural urbanism is crucial and necessary to include diversity as its socioeconomic components.

Urbanism has been historically the outcome and materialization of the technological mode of production in various forms such as transportation, the built environment, energy supply, sewage and water treatment, and industrial production. This includes the issues, constraints, challenges, and opportunities of technologically globalizing connectivity of information in relation to work, live, and play in an urban context. In the Age of Global Information, the gap between rural and urban area has been diminished by the internet connectivity. The global connectivity has partly demolished the borders between geographical places and the legal construct of territory. Anthony Gidden identifies this global connectivity as the consequence of modernity; this connectivity intensifies worldwide social relations that connect geographically distant localities in such a way so that local happenings are shaped by the events occurring of kilometers away and vice versa (Gidden 1990, 64).

For urbanism, global connectivity challenges the definition and inscription of territoriality for the public sphere. The consequence of this global connectivity is the crisis of the public sphere that brings about the dynamics of power relationships between the hegemony of the global capitalist economy and the territorial governmentality of the nation-state. All over, public sphere in the urban settlement is not a geographically and legally constructed territory of nation-state sovereignty, but a conceptually constructed realm of open interactions and communication for all urban citizens; this conceptual territory is challenged by the global connectivity of cyber access and network for its inclusivity and transparency.

To what extent the nation-state has the right to control and manage the cyber interaction and communication that is in alignment with civil liberty and human rights? The Age of Global Information puts the concept of the public sphere in the crisis of its conceptually defining territory and its legally inscribing boundary. The problem of this redefining and re-inscribing lies in the disposition of the public sphere as a human condition for urbanity. There are two essential components of the public sphere in the global urbanism, namely capital and information. Theoretically speaking, both components work for the global economy under the framework of civil liberty. However, globally democratic governmentality seems not—under any circumstance and reason—for being able to manage the global capital network and worldwide access to information.

Moreover, the media in the Global Age are not impartial from the expansion of global capitalistic marketing, production, and distribution. Cable television programs, billboards, and internet pages are never free from commercial advertisements of global products and services. The bargaining power of global corporations is immense that enable them to penetrate into geographically distant locations from New York, Tokyo, London, Paris, and Singapore. However, people from those places are instantly able to see and feel the features of these world metropolitans; all these are experienced through the commercial broadcastings of things and events that matter for the modern lifestyle. The sensual imposition of this lifestyle to local people has been working effectively with entertaining advertisements; their messages are conveyed with massive bombardments of audio-visually luring scenes on the digital screen and other media. In many ways, global entertaining and reporting media do not only fill the gap between local people and global commodities but also transform local perceptions from geographically attached tradition to globally exposed habitation. The use of wireless gadgets and devices changes the way how people understand the concepts and values of privacy, intimacy, control, and respect; in most cases, people can run from their affairs but cannot hide their whereabouts. Internet-based social media become the sites where such transformations take place. For urbanism, global social media unfold, enrich, and accelerate the concept of the public realm on a virtual platform. Despite their detrimental excesses for personal life and reputation, global social media bring about people together with a new sense of proximity and connectivity.

Global Urbanism and Man-Made Environment

Global urbanism is represented by architectural signs and symbols of worldwide corporations. Such representations are visually recognizable from their iconic brands and buildings. Architecturally speaking, most contemporary cities on the globe contain and perform the commercial billboards, advertisements, and broadcasts of internationally operating businesses—from petroleum industries, high-technological computing and telecommunication manufacturers, automobile producers, globally franchised gas convenience stores, stations, shops, groceries, restaurants, and cafes, to internationally well-known enterprises for apparels and appliances—characterize the main streetscapes of most cities on the globe today.

To a certain extent, there is hard to find quite the main streets on the globe, which are free from the manifold advertisements, signage, attributes, and labels of worldwide commercial brands and products. Despite their often-noisy appearances, signs, and symbols of global companies uphold the vibrancy and animation of urban streetscapes and facades. Where ever people go around the globe, they will find similar and recognizable signs and symbols of the products and services they are familiar with, have seen and known before. The contemporary cities are overflowed with various things and amenities, which are available everywhere as the consequence and logic of global capitalist way of production and distribution.

The familiarity with the products and services of global enterprises give people the sense of emotional security where ever cities they visit and stay; this experience is mostly valid from the perspective of visitors from Global North to cities in the Southern Hemisphere. The signage of Starbucks, McDonald, Burger King, H & M, Philips, Samsung, Toyota, etc. gives people the impression that the cities they visit are in the hub with theirs. However, for urban dwellers of cities in the Global South, the signs, symbols, and attributes of worldwide representatives of multinational corporations can be considered as the visually dominating regime of the overseas country over indigenous visual culture. At this point, global urbanism is visually experienced as an architecturally colonializing fashion and style that diminishes and degrades the connectivity of people with their locality.

In response to the flux of globally franchised businesses, most local municipalities have neither a long-term anticipatory strategy nor direct regulatory plan. The common principle for managing global signage and attributes is to apply public safety for the construction and operation within the building codes. Aesthetically speaking, the domination of global signs and symbols at streetscape remain bold and obvious. To a certain extent, this visual colonialization is dramatically staged with the installment of gigantic digital screens covering the façade of historical buildings and public places. On such screens, the pulse and vibrancy of the outside worlds are brought into the city as an integrated part of local livability.

One important contribution of global urbanism in the Global North to the liveability of a city is the enrichment of its streetscapes with a variety of amenities from different places on the globe. Strolling along the streets become enjoyable and attractive. This case is made possible with the urban design policy and land use planning framework that are supportive of the pedestrian-friendly environment with transit-oriented development.

In the Global North, urbanism has been able to transform some streetscapes from monotonous and uniform architectural experience to diverse and complex activities with multicultural amenities. Such multicultural diversity brings about a new horizon of acculturation and syncretic cultivation for new things and events; culinary, arts, and performances are most benefited from multicultural society. However, in sociopolitical areas, the integration of immigrant culture in the mainstream habitation is not always smooth. Adjustment, tolerance, and negotiation become a daily routine in public spaces. In this process, formal education plays an important role in the process of sociocultural integration of immigrants into the mainstream of the Global North societies. Immigrant parents learn and have to negotiate with their children who are going to schools with different cultural values.

Concluding Notes

Urbanism is not about the exclusivity of the daily life-world but the experience of being included in the global network and platform that brings about the sense of interactive proximity and borderless global society. This is the state of affairs of global urbanism in which the territorially administrated authorities are challenged to deal with the globally flowing capitals, transnational public safety, global public health, and cosmopolitan citizenship. The global urbanism is a twofold subject that is associated with the danger of homelessness and the opportunity of remaking our understanding of urbanism in the Age of Global Information. Whatever the danger is, the essential mission of urbanism for civility based on proximity and social orderings system should prevail through the inclusivity of diverse ideological positions, political economic approaches, and cultural values.

The urban settlement does not make any sense without its intentionality for building a home for the urban community. Therefore, the advent of writing is not sufficient for the establishment of urban community and settlement. There is still the necessary governing body that keeps the integrity of place and the urban life-world. This governing agency is essential for managing the dynamics within the human community for their existence, transformations, and developments. The importance of governmentality is not simply for the stewardship of social order within the urban area. Rather, the governmentality is to manage power relation for cultural production and reproduction in relation to other cities and regions.

The building of a home provides the spatial and cultural boundary of the life-world. Urban fabric builds the spatial and geographical boundary that provides people with the feeling of safe and secure from the infinity of nature. While scribe and norms give people with the sense of familiarity for self-expression as well as for communication and interaction with others in a spontaneous way. The sense of home in urbanism has been shown in history that it is possible because there is a sociopolitical ordering system in terms of laws, customs, and norms. Such a system has been historically associated with the existence and sustenance of government that represents the steward and authority of the social order for the urban life-world. The exigency of the governmental institution comes about in response to the call for the guardian of social order and public safety. Since the world is the understandably interrelated system of beings based on language, relations among beings are necessarily organized and maintained within the framework of the sociopolitically ordering system.

In urban context, civility and urbanity are interchangeable concepts; both are about the intentionality of human being that attracts, gather, mobilize, and direct human works, labors, and actions toward human highest disposition as a political being. Regarding its directedness for human activities, there is nothing more striking for urbanism than the concept of urbanity and civility that take place in urban territoriality; this is what humankind strives and pursues for and why they build, develop, uphold, and sustain their urban settlement in the form of town, city, and metropolitan. In a matter of fact, urbanity and civility are historically inseparable from the urban life-world affairs that manifest in various forms of interaction and exchange. Both notions are related to each other in order to guard the sustainability of economic and cultural production.

References

Abrahamson, Mark. 2004. *Global Cities*. Oxford: Oxford University Press.
Acuto, Michele. 2013. *Global Cities, Governance and Diplomacy: The Urban Link*. London and New York: Routledge.

Amen, Michael Mark, Kevin Archer, and Martin M. Bosman. 2006. *Relocating Global Cities: From the Center to the Margins.* Lanham: Rowman & Littlefield.

Arendt, Hannah. 1958/2013. *Human Condition.* Chicago: University of Chicago Press.

BBC World Service. 2016. "www.globescan.com." April 27. Accessed April 28, 2016. http://www.globescan.com/images/images/pressreleases/BBC2016-Identity/BBC_GlobeScan_Identity_Season_Press_Release_April%2026.pdf.

Birch, Eugenie L., and Susan Wachter. 2011. *Global Urbanization.* Philadelphia: University of Pennsylvania Press.

Bishop, Ryan, John Philips, and Wei Yeo Wei. 2003. *Postcolonial Urbanism: Southeast Asian Cities and Global Processes.* London and New York: Routledge.

Bourdieu, Pierre. 1986. "The Form of Capital." In *Handbook of Theory and Research for the Sociology of Education,* edited by J. Richardson, translated by R. Nice, 241–58. New York: Greenwood.

Bourdieau, Pierre, and John B. Thompson. 1991. *Language and Symbolic Power.* Cambridge, MA: Harvard University Press.

Brenner, Neil, and Roger Keil. 2006. *The Global Cities Reader.* London and New York: Routledge.

Calthorpe, Peter. 2010. *Urbanism in the Age of Climate Change.* Washington, DC: Island Press.

Castells, Mario. 1996. *The Rise of the Network Society.* Oxford: Oxford Blackwell.

Dawson, Ashley, and Brent Hayes Edwards. 2004. *Global Cities of the South.* Durham: Duke University Press.

Delanty, Gerald. 2000. *Citizenship in a Global Age: Society Culture Politics.* Buckingham and Philadelphia: Open University Press.

Florida, Richard. 2008. *Who's Your City? How the Creative Economy Is Making Where to Live the Most Important Decision of Your Life.* Toronto: Vintage Canada.

Gidden, Anthony. 1990. *The Consequences of Modernity.* Stanford: Stanford University Press.

———. 1998. *The Third Way: The Renewal of Social Democracy.* Cambridge, UK: Polity Press.

Grimley, Naomi, and BBC. 2016. "Identity 2016: 'Global Citizenship' Rising, Poll Suggests." April 28. Accessed April 28, 2016. http://www.bbc.com/news/world-36139904.

Gugler, Josef. 2004. *World Cities Beyond the West: Globalization, Development and Inequality.* Cambridge, UK: Cambridge University Press.

Hall, Peter Geoffrey. 1998. *Cities in Civilization.* London: Pantheon Books.

Harvey, David. 1998. *The Urban Experience.* Oxford: Basil Blackwell.

———. 2008. "The Right to the City." January. Accessed March 1, 2016. http://davidharvey.org/media/righttothecity.pdf.

Heidegger, Martin. 1977. *Basic Writings.* Edited by David Farrell Krell. Translated by Adolf Hofstadter. San Francisco: Harper and Row.

Huntington, Samuel P. 1993. "The Clash of Civilization?" *Foreign Affairs* 72 (3): 21–49.

Kennedy, Roger. 2014. *The Psychic Home: Psychoanalysis, Consciousness and the Human Soul.* London and New York: Routledge.

King, Anthony D. 2015. *Global Cities.* Routledge Library Editions: Economic Geography. London and New York: Routledge.

Krause, Linda, and Petrice Petro. 2003. *Cinema, Architecture, and Urbanism in Digital Age.* New Brunswick, NJ and London: Rutdgers University Press.

Lechner, Frank J., and John Boli. 2012. *The Globalization Reader.* Malden, MA: Wiley-Blackwell.

Levebre, Henri. 1991. *The Production of Space.* Translated by Donald Nicholson-Smith. Malden: Blackwell.

Miraftab, Faranak, and Neema Kudva. 2015. *Cities of the Global South Reader.* London and New York: Routledge.

North, Douglas. 1991. "Institutions." *Journal of Economic Perspectives* 5 (1): 97–112.

Olson, Kevin. 2008. "Governmental Rationality and Popular Sovereignty." In *No Social Science Without Critical Theory,* edited by Harry F. Dahms, vol. 25, 329–52. Bingley: Emerald.

Parnell, Susan, and Sophie Oldfield. 2014. *The Routledge Handbook on Cities of the Global South.* London and New York: Routledge.

Robertson, Roland, and Kathleen E. White. 2003. *Globalization: Critical Concepts in Sociology.* London and New York: Routledge.

Roy, Ananya, and Nezar AlSayad. 2004. *Urban Informality: Transnational Perspectives from Middle East, Latin America and South Asia.* Lanham: Lexington Books.

Roy, Ananya, and Jennifer Robinson. 2015. "Global Urbanism and the Nature of Theory." *International Journal* (Willey) 40: 1–11.

Sassen, Saskia. 1991/2013. *The Global City: New York, London, Tokyo.* New York: Princeton University Press.

Scott, Alen J., and Michael Storper. 2015. "The Nature of the Cities: The Scope and Limit of Urban Theory." *International Journal of Urban and Regional Research* 39 (1): 1–15.

Sheppard, Eric, Helga Leitner, and Anant Maringanti. 2013. "Provincializing Global Urbanism: A Manifesto." *Journal of Urban Geography* 34: 1–9.

Short, John Rennie. 2014. *Urban Theory: A Critical Assessment.* London: Palgrave.

Xiangming, Chen, and Ahmed Kanna. 2012. *Rethinking Global Urbanism: Comparative Insights from Secondary Cities.* London and New York: Routledge.

Part II
Empirical Exploration

Fig. 1 Urban Scene of Old Jakarta 2015. Photograph by author

8

Everyday Life of Urbanism in the West Malay World

Everyday Urbanism

This chapter is a study of daily urbanism in the West Malaysian world. The focus of the inquiry is the search for manifold aspects of what urban settlement in this region is. In dealing with the question, this study will interrogate the local concepts concerning organization, structure, and function of urban settlement. The interrogation of concepts is directed to dismantle and unfold the content of understanding of places and happenings that bring about a socially constructed reality we understand it as the life-world (Husserl 1970; Bennet 2005, 19–37); the realm in which people experience their socially shared values and their societal identity. The life-world in this study is understood as a whole system of things and affairs that involve people and domains on regular basis. Accordingly, town or city might be a concept which is drawn and mathematically constructed from the practical experience of the life-world. The life-world is a socioculturally constructed reality that contains unique and customary values system (Berger and Luckmann 2011). The urban life-world exists in a way of habitual involvements and engagements of people in the public realms.

© The Author(s) 2020
B. Wiryomartono, *Livability and Sustainability of Urbanism*,
https://doi.org/10.1007/978-981-13-8972-6_8

Even though urban settlements are and do exist as a system, their phenomena are barely definable (Smith 2007, 10). There is hardly to come into consensus what urban settlements are? So what the matter with the facts that despite urban citizens may not necessarily know each other they are bound up together as a whole; regardless of their background, beliefs, occupation, social milieu, and income, the sense of belonging together as a nation is experienced and 'is' as presence; such a phenomenon is coined by Anderson (2006) as an imagined political community. The question is: how such an imagined community exists in daily life? This study attempts to find out the constituent events of the daily life-world that attract and gather people together as urban dwellers. To what extent does the imagined community work on a daily basis in Malaysia?

The urban settlement is considered a concept of an imagined political community; its existence is based on the practical experience of what 'is' and has been performed by its citizens. Urbanism exists in the domains and forms of a politically organized boundary. Town or city is politically a nation because it is not like a village community; the membership of urban settlement is not based on a practical tie but on that which is imagined by its members. As a whole system, the urban settlement is and performs habitations and events that bring about into the light its relation to, the function of, and association with an imagined community. The daily life of urbanism is the phenomena of the life-world based on an imagined community; the phenomena manifest either as the daily practical world or as the occasionally unique events of urban inhabitants. Nevertheless, the urban community's life-world is not necessarily established on the basis knowing each other based on face-to-face interactions, but by the people's imagination as a part of such community, inherently limited and autonomous (Anderson 2006).

Under the notion of an imagined community, town or city does exist as a home of people based on their collective memory (Klimasmith 2005, 4). The urban settlement is likely hard to survive and develop without its sociocultural and economic activities. As a collective home place, town or city needs to run the economic household of urbanism. Even though the urban household is complex and multifaceted, all citizens are likely united with one concept: an imagined urban community. The question is how such an urban settlement recognizable as a community if it does not perform? This study is an attempt to deal with the question of identity for an urban settlement based on the daily performance of urban citizens. Accordingly, connecting the dots of relationship between people and domains is crucial in understanding the aspects of urbanism that matter for an imagined socioeconomic and political community. Without this demonstrative relationship, an imagined community is barely proven as a working concept for urbanism. Methodologically speaking, the relationship will be identified with participant observation on sites; then data of observation will be examined and scrutinized according to spatiotemporal categories to find a pattern in the context of habitual facts.

What Is Malay Urban Habitation?

As collectively memorial achievements, the sense of urban habitation is likely inseparable from the search for a home in the context of society. To a certain extent, the concept of habitation is associated with the notion of culture or civilization because sociocultural cultivation needs a stable and resourceful settlement (Atkin and Rykwert 2005). Even though there are fine and slight differences between 'culture' and 'habitation' on their values, this study tries to manage both concepts in a similar and equivalent level regarding their practices. However, habitation shows and demonstrates how culture does and works in a society. Since as a set of human endeavors, culture is a textual system written by native people (Geertz 1983, 50); local concepts are an essential source of its knowledge.

Urbanity and locality are indivisible in terms of the social construction of the spatial and spatial construction of the social (Lefebvre 1991; Massey 1994); both concepts constitute a local narrative of hometown. Such a narrative is established and cultivated by their regular gatherings in the public realm; such gatherings involve people, places, and things in various architecturally built forms and local customs and traditions. However, as collective work, urbanity is actually the product and process of socioeconomic interactions that involve assimilations, adjustments, negotiations, and accommodations. The question is: what are the sense and essence of habitation in a socioeconomically concentrated place and highly populated settlement? In dealing with the question, it is necessary to ponder the original concept of what urban is. Etymologically speaking, urban is from Latin *urbs* or *urbanus* meaning derivatively civilized, polished, and refined behavior (Partridge and Partridge 1977, 3636). Being civilized with polished behaviors and refined manners is nothing but the quality of being *urbane*. Urbanity is likely not simply a product of human interactions. Rather, urbanity is actually a learning process and sustainable cultivation toward a fine and sophisticated urban society, regardless its size and geography. Urbanity includes the daily learning process of refinement and complexity for being urban society.

Since the sense of habitation lies in its daily life of capacity and practical action, its happenings in spatiotemporal setting and context show and demonstrate recurrence and disposition, as coined by Bourdieu in terms of *habitus*. Based on these recurrences and dispositions, one is able to construct the socioculturally imagined realm of home place of urban society. Pierre Bourdieu's studies (1984, 1986, 1990) show that *habitus* is a set of dispositions; such dispositions engender and generate social practices, perceptions, and behaviors of people; their behaviors are ordinary without consciously regulated by any rule but by simply by practically repeated and affirmed repertoires. Such repertoires establish unconscious and spontaneous dispositions of *habitus*. Habitation is a system of social practices of *habitus* that comprise the incorporations of capitals, fields, and conducts in a socially built environment called habitat. Accordingly, the socially ordering system is inscribed in the design of the built environment in terms of domains and settlements that corresponds and is consistent with the social norms and values of society. *Habitus* predisposes the members of a society to behave and interact in compliance with the social norms and values of their group.

Regarding the fact that *habitus* is socially historical and spatiotemporally enduring process that is based on the accumulation and cultivation of local interactions, adjustments, and accommodations, its practices establish the sense of home for the people who occupy and inhabit them: domains and settlements; the habitat or the socially built environment provides people with the social map as well as mnemonic device of social norms and orders of their dispositions; the dispositions comprise gestures, postures, ways of speech, seat arrangements, ways of addressing others, and ways of occupying spaces and composing forms. On the other way as well, dispositions prescribe and inscribe the design of the socially built environment with unique concepts, forms, domains, features, and treatments. Methodologically, all concepts, forms, domains, features, and treatments related to and associated with the dispositions of urbanity will be scrutinized and examined to find their relationships that uphold, develop, and sustain the reality of urbanism in the Malaysian world.

The Intentionality of Malaysian Habitation

What is the intentionality of urban habitation? Intentionality is the property of mental states and activities by which they are about or directed at or of things and states of affairs in the life-world (Searle 1983, 1). Intentionality navigates and directs habitation toward specific activities: productions, services, and leisure. In the Malaysian world, the intentionality of habitation is generated from the necessity for social and economic as well as political interactions. Collectivist Malaysian tradition underlies the necessity for this interaction (Hofstede 2001, 234; Abdullah 1996). However, this necessity is indescribable without thinking of habitation as something that upholds the local sustainability of economy and social identity. Urban habitation in the Malay Peninsular world is an outcome and process of the dynamic system of multiethnic interactions which are considered a historical process of amalgamation, assimilation, adjustment, and acculturation. Historically speaking, the Malaysian Peninsula is not only the site of melting pot between Western and Eastern civilization for the exchange of spices (Morrison 2002, 122), but also the active player of syncretism, assimilation, and coalition of various cultures and traditions of Southeast Asia—the natives of archipelago, Siam, Indo-China, China, and India.

What is urbanity in Malaysian urbanism? In the daily context of the Malaysian world, urban culture, *budaya bandar*, is experienced as an outcome of a socially transmitted system of urban productions and services called *perbandaran*. The most important thing about this *perbandaran* is in the form of customary laws, *adat*. The institution that establishes keeps develops and sustains such customary laws is *majelis perbandaran*. The integrity of the customary laws lies in the existence of its *majelis* meaning assembly, meeting, and formal deliberation. The modern institutional form of *majelis perbandaran* is understood as municipality or township. Unlike village or kampong, *bandar* as an urban settlement in the Malaysian world is a political community with three main public bodies: governance (*kerajaan*), religious-spiritual establishment (*majelis agama Islam*), and economic institution (*pekan/pasar awam*). These three institutions have been surviving and developing as the pillars of urbanism in the Malaysian world since pre-colonial times.

Compact urban built form and reliable communication and transportation are part of an urban culture which is conducive for sustainable urbanity in terms of *perbandaran*. However, proximity and diversity need to be considered as the essential components of the urban environment for the creative urban community because both categories stipulate cultural productions in an efficient way. Nonetheless, a well-organized system of production and distribution, reliable infrastructure, and availability of raw materials are important factors for this community as well. The creative urban community is the sign of dynamic urbanism (Ellin 2013, 2) that is committed to the excellence of being *urbane*. In daily habitation, casual interactions are to demonstrate *budaya* in terms of hospitality, *ramah-tamah*, and courtesy, *sopan-santun*. This includes greeting to each other by nodding and smiling when they encounter face-to-face, regardless either they know or do not know each other. In the Malaysian tradition, the concept of otherness is not culturally conceived of as a foreigner or stranger with suspicious connotation. Rather, they are perceived as newcomer or guest who deserves respect and dignity. They address a foreigner as *orang asing* or *pendatang*, in the connotative tone of the not yet known person.

To what extent do locality and urban habitation belong together? Nevertheless, without locality, habitation is hardly animated as something unique and particular as well as memorable and special. This implies that the meaning of locality is another word for the sense of home, which is communicatively animated by means of local history and narrative. In this sense, the spoken language in town is not simply the means of communication. Rather, it is the sign and representation of local urbanity. In the context of Malaysian urban habitation, language is the stockroom of what people have cultivated from their daily life-world in various forms of slang (*slanga*), sonnet (*pantun*), and aphorism or idiom (*perumpamaan*).

Any outcome of cultural invention and experience is subject to identifica-tion, signification, and classification within the system of language. The relationship between urban culture and the local community is character-istically established and developed with the sense of home that comprises locally spoken language, gastronomy, customs, and socially shared values; all this is about locality, *tempatan*. The relation of person to locality is performed with the habitus of *balek kampong*.

The central idea of people going back home, *balek kampong*, is to perform respect and humility to their mother. The Malaysian concept of home is signified by the care and nature of motherhood. The importance of motherhood is incorporated in the design of the traditional Malaysian house that the whole inside domain is dedicated to mother, *rumah ibu* (Tadjuddin et al. 2005, 36). Mother, *ibu*, is the center of the Malaysian universe in a way that is not simply by its feminine nature but more about its spacious characteristics. It is the reason why headquarters and capital cities are called with *ibu pejabat* and *ibukota*; it does not mean that mother is a leader, but it is the place where the leadership, guidance, control, and management come from. Motherhood is also the reference of hospitality and courtesy. To be a Malaysian person is to be a proper child who is respected and follow the mother's guidance toward a person with fine manners, *berbudi bahasa halus*. The sense of human cultivation for the Malaysian culture goes back to the concept of mother's education at home.

In the Malaysian speaking world, the concept of human dignity as well as urbanity refers to the Arabic notion of *adab* meaning well-mannered behaviors, prudent, cultivated, and mature. Originally and the pre-Islamic root of the concept describes *adab* as a habit or a practical norm of conducts that connotes praiseworthy and inherited from tradition, but in early Islamic civilization, the word '*adab*' means similar to the Latin '*urbanitas*' encompassing etiquette, civility, and urban refinement (Fabos 2008, 98).

The Malaysian urbanism adopts *adab* in terms of respectful, polite, and kind behavior. In the practical sense of daily life, this includes nodding to each other and exchanging smiles with fellow citizens, regardless they know or do not know each other, are part of this notion of *adab*. Under the Islamic influence, gender plays an important role in the encounter. However, starring, hugging, and shaking hands by different genders are considerably not appropriate (see also Alhady 1965; Shariff 2004). Since being *urbane* in the Malaysian context is known as *beradab* which is indispensable to the cultivation, *perdaban*, of all *habitus* that includes the regeneration, improvement, and enhancement, as well as replenishment and refinement of habitation. Intense interactions with Muslim trades, Chinese merchants, and Indian mercantile people have been the generator of this cultivation since the fifteenth century. All this made possible because the Malaysian people believe in the culture as *budaya* that will get prevailing and strong when it encounters and interacts with foreign elements and influences.

Nevertheless, the accelerator of urban cultivation is diversity and proximity that stipulate a supportive environment for development and elaboration. Living in *bandar* is characterized by the condition of living in proximity, *duduk berdekatan* meaning literally sitting close to each other. Idiomatically, in the Malaysian world, sitting is a concept used for dwelling and living. The notion of sitting becomes important in Malaysian society because having a seat has something to do with social status, pride, and authority. The sign of this social status is denoted and designated by a patrimonial aristocratic title before their names such as Datuk Seri, Tunku, and Datuk. In an urban area, social status plays an important role in Malaysian society. The sociopolitical establishment determines the direction and priority of development.

Because of nearness, habitation is made possible. The Malay concept for such a nearness is *duduk* meaning sitting and settlement. Having a seat enables one to find a place in society. In Malay culture, having settled down is necessarily having a seat in society with a respectful position, *darjat*, and pride, *martabat*. In the Malaysian world, sitting in proximity, *duduk berdekatan*, is the underpinning of habitation—*tempat duduk*; this concept includes norms and principles of all behaviors, perceptions, communications, expressions, and articulations according to local community's tradition. In short, *adat istiadat* is a system of customary laws which is handed down for generations with oral tradition. Habitation in the context of customary law, *adat istiadat*, is distinguished from habit, *kebiasaan*, because of its social and moral consequences.

Diversity creates and urges various demands of idea, taste, and interest. With diversity, urban community is challenged and compelled to offer more and better products and services as well as actions and occasions. The Malaysian language understands diversity in terms of various modes and appearances; it is conceived of as the enrichment and enhancement of a unified whole. The patrimonial culture of Malaysian society urges and maintains unity and harmony of various elements as the basic condition of the safe and healthy nation-state (see also Case 1995, 84). The roots of this patrimonial tradition dated back to the ancient Malaysian communities based on kampong polity. Seniority and well-experienced leadership play an important role in kampong society; its pervasive power relation is patrimonial comprising the role of father, *bapak*, and son, *anak-buah*. Disloyalty, demotion, and opposition against the father are perceived not only as disrespect to the person but also as an attack, sickness, and disease of the whole society. Therefore, culturally, politically, and societally, diversity is painful and dangerous if it is not spiritually unified.

In Malaysian political culture, the unity of diversity, *perpaduan* is the most primary condition. The root of the concept of *perpaduan* is from the word '*adu*' meaning compete, face to face, race, and participation. At urban and state performance, the concept of *perpaduan* becomes crucial in a way of calling for care and healing of society. For this purpose, urbanism in the Malaysian world values and signifies state rituals as sacred occasions; the events are not only for celebrations but actually the call for unity and harmony. The political leader will do his role and function as the priest of the state deliberating speech with patronizing, prophetic, and messianic tone.

In the Malaysian world, culture is understood as *budaya*, derived from Sanskrit loanword meaning values, works and way of life; the sense of the word entails harmony and unity in society. *Budaya* stems from the words *budi* and *daya*; *budi* means well-mannered, subtlety, cultivated, and mature while *daya* means power, force, and drive. In the broadest sense of the word, *budaya* is the source and power of human dignity based on Malaysian customs and traditions. A good citizen is signified with the word *budiman* meaning a person with gentleperson quality because he/she has *budaya* in a practical sense. All this values system is preserved in the Malaysian language in the form of *pantun*, prosaic lyrics, *gurindam*, poems, and *pepatah*, proverbs and idioms. In daily life, the Malaysian people practice and exercise their *budaya* in terms of *habitus—adat istiadat*.

The question concerning habitation, *kebiasaan*, is undeniably challenging in regard to the boundary of the Malaysian language. The obstacle of habitation lies in the fact of its multi-ethnicity consisting of the Indian, Chinese, and Malay group. Each ethnic group speaks their native language. In the Malaysian context, habitation is conceived of as the act of developing all human faculties and skills that establish, develop, enrich, and sustain the relationship between people and their place. The thing of urban habitation in the Malaysian world claims its existence in the propensity and exigency of the proximate and diverse society. Concerning social order, the Malaysian ethnic group is provided with customary laws, *hukum adat*, which are not documented with the written form, but they have been working well for generations regulating land tenure, inheritance, minor criminal law, and family law.

Even though historically, there was no a unifying system that brings all ethnic groups into a whole system of language, Islamic polities of the Sultanate state had been working to provide social order. Prior to the arrival of Westerners, the Malaysian Sultanate states took care for the security and peace in the region by relying on the good relation with neighboring states: Aceh and Thailand, and of course the Chinese Emperor. The decline of the Chinese Emperor in the sixteenth century was also the weakening support for peace and security in the region. The Portuguese sea power controlled the region with its culmination at the fall of Malacca Sultanate in 1511. Later, the Dutch VOC took over the control over the Straits until 1841 when the British colonial rule undertook the control over the Straits of Malacca as the consequence of Napoleon's War defeat in Europe. Since then, the British common law legal system was put into effect for all people under its jurisdiction in the Malaysian Peninsula.

Unlike the Dutch colonial power in the Indonesian archipelago, the relationship between the British authority and the Malaysian Sultanates was not based upon direct domination and control but was subtly arranged with treaties and agreements. However, in the treaties and agreements were clearly regulated and disposed of that the Malaysian Sultanates were under the direction of the British rule for the economy and political power. The British interest in the Malaysian Peninsula was not impartial from the need for natural resources for the domestic and European market, such as tin ore, spices, and agricultural products. For this purpose, the British authority arranged with the local Sultanates for opening mining and agricultural concessions. In order to keep peace and development for urbanism, the British patrimonial authority granted the indigenous rulers the space for religious and customary affairs of indigenous populations, while the access and control for economic resources remained at the hand of the colonial master.

The European investors and Chinese groups were managed by the British authority to run the mining industries in various places in the late of the nineteenth century until the beginning of the twentieth century. These were the times when the establishment of mining and agricultural towns took place in various states of the Malaysian Peninsula. Chinese migrants came to the peninsula mostly for mining jobs and trades while most Indian Tamil people were brought by the British companies from their homeland to the Peninsula for working in the plantations.

Even though the Malaysian, Chinese, and the Malaysian Tamil children do not go to the same schools, the official language at public service is the Malaysian language (Jomo and Wong 2008, 15). Since the British colonial era, racial segregation for school system has been not resolved and dissolved until today. Urban habitation in the Malaysian towns and cities has to deal with racial prejudice. In the street life, the three ethnic groups live and work together as urban citizens in various fields. What makes them different is actually the government policy. The 1957 Constitution gives the Malaysian ethnic group privilege for most fields of modernity comprising public education, business permits, loans, and government jobs (Di Piazza 2006, 31).

The Features of Malaysian Urban Habitation

In order to understand Malaysian urbanity, we can see urban culture as a system of products and processes which are required and done by urban citizens. All these are established, developed, elaborated, and sustained by urban needs for living, working, and playing which are characterized with proximity, specialty, creativity, diversity, complexity, productivity, actuality, and social mobility. The forces that construct urbanity in the

Malay world are mostly driven by the lifestyle in the logic of economic productions and exchanges and the collective necessity for spirituality.

Activities which are affiliated with mosque, church, temple, and school are the daily life of urbanism based on the collective need for culture and spirituality. The Malay patrimonial culture keeps, upholds, and sustains this daily life of activities within the framework of formal and ritual organization and habitation. Casual interactions for social intimacy are human conditions for their growth and development as members of urban society. Beyond formality, urban inhabitants in the Malay Peninsular world urge and strive to actualize their informal and casual interactions and gatherings that fall into the category of daily life of urbanism. The following places of interactions and gatherings are potentially developed through environmental design for place-making intention and program.

Pasar Awam

Since ancient times, the need for being part of the crowd has been manifested in the phenomenon of the market. From ancient Babylon bazaar to the European Middle Age farm market and modern shopping mall, the sensibility and pleasure of casual gatherings and exchanges have been shaping and cultivating urbanism, *perbandaran*. The Malaysian urban areas and non-urban ones are distinguished by the existence of the public realm for exchanges, *tempat awam*. The heart of urbanism in the Malaysian world is the livability of the public realm at the intersection of streets or at the estuary of the river. As the reality of urbanism, *Pasar Awam* is the place of happenings that casual and formal occasions take place; it is the place of equality in which social status, privileges, and attributes dissolve into an ordinarily depersonalized sphere. This experience brings about the sense of freedom in action. The phenomenon of *pasar awam* does not likely come out from fear of crime, poverty, homelessness, and discrimination, but from freedom and casualness.

In the Malaysian world, marketplace, *pasar* or *pekan*, is the gathering place for social and economic exchanges. The need for mere exchange of commodities is obviously not only the reason for going to the marketplace; most people go to this place simply to get into the crowd. In the Malaysian tradition, the original meaning of the word *pekan* is: gathering, meeting, and have something together that happens once a week. The happening takes place in a permanent location in which people barter and exchange their goods and foods. As a unit of duration for a week, *pekan* consists of seven days: *ahad, isnin, selasa, rabu, kamis, jumaat,* and *sabtu.* After the embrace of Islamic teaching in the fourteenth century, most Malaysian traditional markets were gradually transformed from weekly *pekan* to daily *pasar;* the concept of *pasar* is elaborated and borrowed from the Persian *bazaar.* The adoption of the concept *pasar* for marketplace began in coastal vicinities as part of international trades. Today, the words *pekan* and *pasar* are interchangeably in its use and meaning. However, *pekan* connotes the traditional weekly market of towns in hinterland while *pasar* is more about general market anywhere.

By environmental design, the market is more than just a place for the purchasing and sale of provisions and livestock and other commodities. It is the gathering of people where and when trust, fairness, and mutual respect are exercised, cultivated, and sustained on regular basis. All these virtues are considerably the natural fundamentals of urbanity. In the pre-colonial era, Malaysian markets in the coastal areas were managed by port authority called *syahbandar.* Since the Chinese migrants were mostly well experienced in the business, the Malaysian Sultanate gave a Chinese group a concession to establish the port authority that regulated and managed the trades and taxations in the port and its *pasar.*

They are natural because their existence is prior to any law and regulation. The occurrence of the market represents the nature of humanity in their search for communications and interactions based on self-determination and informality. By means of the market, human needs and activities are made possible for being familiar and honest with their feeling and thinking. The forces and drives that work in the market are about fair play and communication. Even though market activities are ostensibly reliant on the demand-supply exchange, the end of these is inescapable from the necessity for being human in terms of collectively sociable condition.

By environmental design, such cultivation and possibilities should be directed by public policy toward good intentionality in terms of pedestrian-friendly environment and nature conservation. Even though

Fig. 8.1 Pasar Awam in Taman Universiti Johor Bahru (a) and Interior of Kipmart (b) (Photos by authors)

the integration of marketplace and public transit terminal have been developed, convenience, accessibility, and public safety of connecting path need improvement, especially from the marketplace to the terminal and vice versa. Domains inside and outside of the marketplace should be able to accommodate families and children. Regarding its tropical setting, the marketplace should provide more shades and cross-ventilating air movements as well as indirect daylight in its market hall (Fig. 8.1).

Pasar Malam

Around the clocks, casual interactions of people sustain the sustainability of urban economy and culture. One of the informal gatherings of people in the Malaysian towns and cities is the Night market of *pasar malam*; this activity happens on a weekly, monthly, or annual basis. Of course, the event happens in the late evening at 3.30 or 4.00 p.m. until night at 10 or 11 p.m. *Pasar malam* takes place along the streets in an inner town or a historic district. The longest street of night market was recorded with 2.4 km Alam in 2012. The origin of the night market in the Southeast Asian archipelago is probably its mild atmosphere outdoor in the late afternoon. Chinese populations enjoy their gathering outdoor during the late afternoon and evening as their long-standing tradition. For Muslims, the tradition of night market is associated with the month of *Ramadhan* when the families prepare their meals for the break of their fasting, then they celebrate the evening after their congregation in the mosque. However, today night market is a daily or weekly event of urban dwellers in Malaysia.

Fig. 8.2 Annual night market on the Jonker Street during Chinese new year celebration in 2011 (**a** and **b**) and Jonker Walk at Hang Tuah Street of Malacca (**c**) (Photos by authors)

Such tradition is also common in most parts of Southeast Asia such as Vietnam, Thailand, Indonesia, and the Philippines.

In most cases, the traditional grand night market is celebrated as the events for commemorating the birthday of the town or city such as in Malacca. In the state of Pahang, Johor, Kedah, and Kelantan, most towns have a weekly night market. The state government or the state night market association regulates the event in their area so that their night market does not happen at the same time and takes a turn on a weekly basis. *Pasar malam* for most Malaysian populations is a lively performing occasion for casual and social interactions with the festive atmosphere and joyful experience for all family members. Mild air temperature and clear sky of the Malay Peninsula attracts people to go out and gather in public space. Of course, foods from various ethnic groups and spontaneous attractions of local artists and performers support the festivity of the night market (Fig. 8.2).

From several experiences visiting *pasar malam* in various towns and cities in Malaysia, the obvious design intervention necessary is to improve the quality and quantity of utilities for health, safety, and convenience. Even though the event of *pasar malam* is mostly temporary, strolling along the street of Night Market needs convenient access from public transit with sufficient lighting and amenities, especially for elderly and children. Until recently, Malaysian people prefer to visit Night Market with private cars. Providing ample parking places becomes an important requirement for the successful event of Night Market. For a sustainable future, automobile reliant regime should be replaced by pedestrian and public transit-oriented

Fig. 8.3 Daily night market in Skudai Johor Bahru 2013 (Photos by authors)

lifestyle. The event should utilize and take place in the place with the already provided infrastructure for public transit. In Bukit Bintang of Kuala Lumpur, such an event has been in place and could be developed in other Malaysian towns and cities (Fig. 8.3).

Mamak Restaurant

Food is always good for people, regardless of whoever they are, as long as they like it then, everything is alright. Restaurant or food stall called *mamak* is one important place where most people mingle and enjoy their food and casual encounter. The name is originally from Indian Tamil word meaning 'uncle'; children usually call the shop holder as 'uncle.' *Mamak* food stall or restaurant usually owns by Indian Muslim; their food is mostly acceptable to Malaysian populations. Since most *mamak* restaurants in Malaysian towns and cities are open for 24 hours, the stalls play a significant role as a melting pot of various races.

The most popular foods from the *mamak* restaurant are *roti canai* and chicken *tandoori* while the beverage is *teh tarik*. In the course of history, *mamak* restaurant serves Malaysian and Chinese food as well such as *nasi lemak*, nasi *goreng kampung, nasi kerabu,* and *mee goreng.* Of course, *mamak*

Fig. 8.4 Mamak Restaurant and 24 hours food stall in Johor Bahru (Photos by authors)

stalls are unique because they serve the similar foods with the Malaysian or Chinese restaurants, but they serve them with specific Indian taste and flavor. I might be right to say that *mamak* restaurant performs the symphony of Malaysian cuisine in one platform by which the best foods of Malaysian, Chinese, and Indian ethnic group are served. The stall is also the gathering place for most Malaysians. Today, most stalls are provided with a big screen for television sports programs, and news. Families and young people usually spend their leisure in the *mamak* restaurant.

By environmental design, the most location of the *mamak* restaurant is located on the street corner or in the middle part of a retail shopping block. To a certain extent, the pedestrian-friendly environment is still a challenging effort for the *mamak* restaurant. The place-making strategy should work out the whole block where the restaurant is located. In a neighborhood—*taman*—*walking* distance within 5 minutes' walk should be preferable for Malaysian populations. For place-making reliant upon public transit, the restaurant should be in a medium-high density neighborhood—above 20 units per acre or 240 people per hectares—with multi-types of homes (Fig. 8.4).

The Chinese Hakka people were well known as a well-organized group for business. In their business, they work together under the concept of *kongsi* meaning collaboration with share-based mutuality. The group was led by a business leader called *kapitan cina*. Famous *kapitan* for Perak was Chung Keng Quee, Hai San for Penang, and Yap Ah Loy for Kuala Lumpur. The Hakka group played an important role in establishing and developing mining towns in various locations of Malaysian, such as Ipoh,

Fig. 8.5 Kopi Tiam in Skudai Johor Bahru (Photos by authors)

Taiping, Seremban, and Bentong. The society of Hakka brought their urban custom and tradition for gathering in public places. *Kopitiam* was originally a male clubhouse in Chinese settlement.

Kopitiam

The other place where urban people in Malaysian towns and cities hang out for leisure or have a good time with friends and family is in a café called *kopitiam*. In Malaysian tradition, such a place is known as *kedai kopi* meaning coffee shop or café. The establishment of *kopitiam* dated back to the Chinese Hakka populations in the Malaysian Peninsula who came from the central mainland of China at the end of the nineteenth century when mining of tin ore in Perak and Penang was explored (Fig. 8.5).

Hanging Out Outdoor

Theoretically, sports are activities that unite people regardless of whoever they are. In most Malaysian urban areas, shophouses and *gelang-gang sukan* are provided in every subdivision or neighborhood, *taman*.

Dining outdoors in the late evening is the scene to see. Various food stalls utilize the parking area or part of the street as their outdoor dining area. Big screen for projected images of the popular television stations is installed to entertain people while dining together with their friends and families.

In the Malaysian context, it is commonly accepted and deeply rooted in their tradition for the sensibility of greeting to each other with respect; there is a pleasure here for casual gatherings by exchanging smiles with each other during the clear weather. Being anonymously present in the crowd watching sports or movie is likely impossible in the neighborhood scale. The waiters or other customers will say hello and want to know each other. Of course, seeing others and being seen by others is part of the Malaysian socialization.

Even though in Malaysian urban areas, the outdoor food court is a common place for social gathering, an ethnic group such as Chinese populations do have their own realm not because they do not want to intermingle with other ethnic groups, but simply because they consume pork and alcoholic beverages in public areas. Nevertheless, the Chinese populations establish and develop specific features and traits that enrich and vibrate the public realm of Malaysian towns and cities with Chinese design elements and their customary events. The annual festival and celebration for Chinese New Year bring about unique and vivacious scenes on the streets and building facades with decorations and lights. Despite ethnic prejudices are often bought up as sociocultural issues by politicians, at the daily life, the Chinese, Indian, Malaysian, and other ethnic populations are able to mingle together. Variety foods have been developed and prepared throughout decades that represent the fusion kinds of food with a unique variety of tastes as well.

The Indian Tamil communities are not left behind in making Malaysian urbanism alive. Deepavali and festival of light are celebrated in many public realms with dramatic attractions, colorful flowers, and festive banners. Even though racial prejudice and tension are not fully dissolved by Malaysian government policies and programs, on the street and daily habitation, tolerance, and friendly interaction are exercised in various fields and occupations. Despite the diversity of races and beliefs, the safe and secure environment and tolerance are probably the conditions of peacefully living together in Malaysian urban areas. However, the roots of the sense of community need further exploration for equal opportunity for all and fairness in all fields of urban habitation. Until recently, the indigenous Malay populations still enjoy the privilege and preference by law for getting government jobs, rewards, investments, and loans.

In the Malaysian world, most cities and towns are the outcome of historical process and development without an intentionally designated plan of the scheme. Only a few of them are the product of modern plan and design such as Putrajaya and Iskandar Malaysia. Several subdivisions, *mukim*, and neighborhoods, *taman*, have been developed with a modern plan and design. Nevertheless, urbanity is likely not determined by formal plan or design, but by the necessities of urbanism such as proximity, connectivity, diversity, and complexity.

Traditionally, the proximity of forms and uses has been part of Malaysian urban settlement with shophouse as its basic element of the urban structure. The two stories row shophouses characterize the urban fabric of most Malaysian towns and cities. The block usually consists of 5–20 shophouses, depending on the area and topography of the site. The Chinese populations, Hokkien, Teochew, Hakka, Hainanese, Heng Hua, Min Dong, Yue, Guang Fu, Sei Yap, Guangxi, and Wu people who emigrated from the mainland have established and developed the hybrid function of house and shop together in various places in the Malaysian world since the fifteenth century. The diverse Chinese populations in terms of beliefs, customs, traditions, and languages contribute to the richness and variety of Chinese cuisines, architectural elements, and fashionable features in the Malaysian world.

Based on the two stories of shophouses, the urban fabric is defined with a linear pattern of shophouse row blocks. The street as a public realm is established between the two structures with an intimate proportion for human scale. Moreover, the flexibility and density in the use of space are well accommodated by the structure of shophouses. So that Indian, Arabian, and Malaysian merchants and business people are able to utilize the shophouse structure for their shops, offices, and other services. All ethnic groups buy or build their own shophouses to integrate their business with the existing structures which mostly have been established by the Chinese populations. In doing so, compact and well-connected urban fabric has been achieved at human scale. It is not surprising to recognize that the historically urbanized areas are populated by the Chinese populations because they have a long-standing tradition for building their architecturally urbanized settlement with a dense, multifunctional, and compact structure by means of shophouse block.

Furthermore, urban architecture encompasses form and space that create and establish places for living, working, and playing. The Chinese row shophouses commonly own by a clan or a *kongsi* that commonly regulates the variety of businesses. *Kongsi* is a concept of cooperative work and business based on the share; several individuals pool their resources together and share the risks as well as the profits fairly (Ooi 2009, 159). In doing so, the diversity and mutuality of businesses and services are economically managed. Chinese populations are also well experienced in setting up and maintaining their business network from upstream to downstream of various fields.

Professional, social, cultural as well as business associations have been a long-standing tradition for Chinese urban society. In every town, the associations have their own clubhouse; membership in such club is not simply a formality, but a life or death choice in the business with the local and global network. It is probably the reason why most fields of businesses in the Malaysian towns and cities are dominated by the Chinese people. The Malaysian and Indian Tamil populations do not have such traditional associations like the Chinese counterpart. In dealing with the Malaysian preference policy, the Chinese business people work together with the Malaysian natives based on *kongsi* concept so that they can grow together with fair risks and profits. Such Malaysian-Chinese collaboration is well known as *Ali-Baba* business.

Concluding Notes

The daily life of urbanism in the Malaysian world is not determined by the concentration of settlement but by the presence of political authority and spiritual center. Sultanate palace and other state buildings represent more about the presence of patrimonial and spiritual authority. The place-making of indigenous urbanism is reliant upon the existence of market-place as the core of the daily life of urbanism. Even though in the Malay

Peninsula, mosque/palace is a potential site for place-making, its formality and regulatory conditions do not allow all urban inhabitants to have social interactions. However, the role and function of mosque/palace and marketplace are complementary in the establishment, development, and sustenance of urbanism in the Malay Peninsular world. Place-making in the context Malay world entails the gatherings of people with various kinds of intentionality, from economic, social, to cultural, and from political to spiritual. The nature of place-making for urbanism involves the daily life of gatherings that establish, develop, and sustain acculturation, assimilation, and expansion of local culture and tradition. Patrimonialism of traditional authority directs, regulates, and controls urbanism toward comportment based on traditional and religious values system, while informal intentionality attracts, leads, and allows people to casual gatherings that build, develop, and sustain corporeally social intimacy among all urban inhabitants; the sense of place-making lies in such gatherings that construct the sense of home.

References

Abdullah, Asma. 1996. *Going Local: Cultural Dimensions in Malaysian Management*. Kuala Lumpur: Malaysian Institute of Management.

Alhady, Alwi. 1965. *Adab Tertib*. Kuala Lumpur: Malay Publication.

Anderson, Benedict. 2006. *Imagined Communities: Reflection on the Origin and Spread of Nationalism*. London and New York: Verso.

Atkin, Tony, and Robert Rykwert. 2005. *Structure and Meaning in Human Settlement*. Philadelphia: University of Pennsylvania Museum of Archeology and Anthropology.

Bennet, Rudolf. 2005. "Husserl's Concept of the World." In *Edmund Husserl Critical Assessments of Leading Philosophers*, edited by Rudolf Bennet, Donn Welton, and Gina Zavota. New York: Routledge.

Berger, Peter, and Thomas Luckmann. 2011. *The Social Construction of Reality: A Treatise in the Sociology of Knowledge*. New York: Open Road.

Bourdieu, Pierre. 1984. *Distinction: A Social Critique of the Judgment of Taste*. London: Routledge.

————. 1986. "The Form of Capital." In *Handbook of Theory and Research for the Sociology of Education*, edited by J. Richardson, translated by R. Nice, 241–58. New York: Greenwood.

————. 1990. "Structures, Habitus, Practices." In *The Logic of Practice*, edited by P. Bourdieu, 52–79. Stanford, CA: Stanford University Press.

Case, William. 1995. "Malay, Aspects and Audiences of Legitimacy." In *Political Legitimacy in Southeast Asia*, edited by Muthiah Alagappa, 69–107. Stanford: Stanford University Press.

Di Piazza, Francesca. 2006. *Malaysia in Picture*. Minneapolis: Visual Geography Series.

Ellin, Nan. 2013. *Good Urbanism: Six Steps to Creating Prosperous Places*. Washington, DC: Island Press.

Fabos, Anita. 2008. *'Brothers' Or Others?: Propriety and Gender for Muslim Arab*. Lanham: Berghahn.

Geertz, Clifford. 1983. *Local Knowledge*. New York: Basic Book.

Hofstede, Geert. 2001. *Culture's Consequences: Comparing Values, Behaviors Institutions, and Organization Across Nations*. London: Sage.

Husserl, Edmund. 1970. *The Crises of European Sciences and Transcendental Phenomenology*. Translated by David Carr. Evanston: Northwestern University.

Jomo, Kwame Sundaram, and Sue Ngan Wong. 2008. *Law Institutions and Malaysian Economic Development*. Singapore: NUS Press.

Klimasmith, Betsy. 2005. *At Home in the City: Urban Domesticity in American Literature and Culture 1850–1930*. Lebanon: University of New Hampshire Press.

Lefebvre, Henri. 1991. *The Production of Space*. Translated by Donald Nicholson-Smith. Malden: Blackwell.

Massey, Doreen. 1994. *Space, Place, and Gender*. Minneapolis: University of Minnesota Press.

Morrison, Kathleen. 2002. "Pepper in the Uphills: Upland and Lowland Exchange and the Intensification of the Spice Trade." In *Forager-Traders in South and Southeast Asia, Longterm Histories*, edited by Kathleen D. Morrison and Laura L. Junker, 105–30. Cambridge: Cambridge University Press.

Ooi, Keat Gin. 2009. *Historical Dictionary of Malaysia*. Lanham: Scarecrow.

Partridge, E. Staff, and Eric Partridge. 1977. *Origin of Etymology Dictionary of Modern English*. 4th ed. London: Routledge.

Searle, John R. 1983. *Intentionality: An Essay in the Philosophy of Mind*. Cambridge: Cambridge University Press.

Shariff, Abbas Mohammad. 2004. *Adab Orang Melayu.* Kuala Lumpur: Alfa Media.

Smith, Peter. 2007. *The Dynamics of Urbanism.* London: Routledge.

Tadjuddin, Mohammad Rasdi, Komarudin Mohammad Ali, Syed Iskandar Syed Arifin, Ra'alah Mohammad, and Gurupiah Mursib. 2005. *The Architectural Heritage of the Malaysian World: Traditional Houses.* Skudai: UTM Press.

9

Urbanism, Society, and Culture in the Malay Peninsular World: Bandar Malacca

Patrimonialism and Urbanism

Ideologically speaking, the practice of patrimonialism in Southeast Asia has been working within society and culture since ancient times. In the local narrative of *Sulalat al-Salatin* or Malay Annals or *Sejarah Melayu*, patrimonialism and feudalism have been inscribed as the political culture of most Malay-Indonesian archipelago since the seventeenth century (Walker 2009; Chambert-Loir 2005). According to the Annals, power is understood as divine or celestial supremacy under the notion of *daulat*. The political hegemony of the ruling group, *orang kaya* or *orang besar,* is conceived of as undisputable because it will lead common people, *orang banyak*, to the errant and sinful state of being in terms of *durhaka*. Under the Malay political culture, the holders of *daulat* become the leaders and landlords—*tuan ku* or *tunku* or *datuk*—while the rest of the populations become their followers, *anak buah*. As a whole, the Malay nation-state is a dualistic system of social groups: the ruling and the ruled. Traditionally speaking, the economic household of the Malay state relies on the patron–client relationship. The patron is at the position of those who have the privileged access to resources and control their uses and productions.

© The Author(s) 2020
B. Wiryomartono, *Livability and Sustainability of Urbanism*,
https://doi.org/10.1007/978-981-13-8972-6_9

Meanwhile, the clients or followers are those who work with or for the patrons in various fields of productions, services, and distributions. As a whole, the concept of egalitarian society is traditionally unknown. The available concept for a group of people is rakyat for commoners and *masyarakat* for the community; however, these concepts do not include the elites or the patron in the Malay system of nation-state, *negara*. The patrons or the ruling people belong to the divines with extraterritorial power and hegemony, *daulat*. Prior to the arrival of Chinese and Indian migrants into the Malay Peninsular world, the social stratification of the Malay state remained dualistic as mentioned above.

Patrimonial culture has been an integrated part of the Southeast Asian tradition, by which the concept of society as a whole is sociopolitically constituted by dualistic components: the ruling group as patron and the ruled as a client. This dualistic stratification system of patron–client ties engenders a categorical status of people with apparent social standing either as the member of elites—orang kaya, orang berada, and orang kuasa—or as that of the commoners—orang banyak, rakyat. In the Malay-speaking world, these categories are recognizable with the concepts of pembesar and *rakyat*. The elite group in Malay populations include the feudal aristocratic families, politicians/members of UMNO party, native business people, and other privileged groups, such as former high-ranking state officials and military functionaries, Indian and Chinese capitalists. The rest of the ruling group belongs to commoners or *orang banyak* that include native people—*bumiputera*—Chinese, Indian, and other ethnic populations; they are the majority in number but do not have a powerful position in multiethnic Malaysian populations.

The concept of society as the socioculturally united system is hardly applicable for this ethnically and economically divided societies, the Malay, Indian, Chinese groups, and other foreigners. From a sociological perspective, post-colonial and colonial social system of Malaysia has not been significantly different. Maintaining the status quo of traditional patronage becomes an important disposition of Malaysian entity as a nation-state consisting of this dualistic sociological group. Economically speaking, the patronage of the elites has been politically maintained by the ruling group. This ruling group has been directly and indirectly associated with the well-established functionaries of UMNO that holds, conserves, develops, and sustains its existence based on exclusive membership among native Malay upper class and other privileged persons.

This study understands patrimonial urbanism as a system of ideas, forces, efforts, circumstances, activities, ways, and values that construct, shape, establish, develop, and sustain settlement and its *urbane* society based on the politically ordering system of the paternalistic regime, either from indigenous or Western colonial origin. As a concept, patrimonial urbanism comes into being as the phenomenon of habitation that manifests in various forms, structures, events, styles, scales, fashions, and expressions. This study is an attempt to dismantle the concept of patrimonial urbanism in Malacca from its historical development and local context. The focus of the study is to find out and examine the traces of the relationship between patrimonialism and place making that enables them to build a multicultural hometown in Malacca. Historian of Southeast Asia, Merle Calvin Ricklefs highlights that the success of pre-colonial Malacca Sultanate's policy was in creating a multicultural community of traders with advantageous facilities (Ricklefs 2001/2008, 24); the Portuguese sea power could take over the port but not able to build such a community after they conquered Malacca in 1511. This study is an attempt to connect the historical dots and traces of multicultural urbanism in Malacca that has not been profoundly excavated, unfolded, and connected as a system of knowledge.

Methodologically, the data of this study are the local concepts, historical notes from literary sources, and the phenomena of urbanism from daily habitation; they are sought and collected from the sites of investigation, *Bandar Melaka* area. In examining the data, this study cross-checks the sources from other cities in the Malay Peninsula such as Muar, Ipoh, Mersing, and Kuantan. This study argues that the urbanism in Malacca works is the exemplary waterfront vicinity that maintains and sustains diverse communities as a system of habitation.

Urbanism and Pre-colonial Bandar of Melaka

The evolution of Melaka from capital to municipality has been intensively studied by Sandhu and Wheatley; their work is considerably classical that provides us with a comprehensive interpretation on Melaka as a unique urban center in Southeast Asia since the 1400s. *Bandar* Melaka is probably one of the best examples of Southeast waterfront city with multicultural populations (Sandhu and Wheatley 1983, 495–597). Melaka was probably the first *entrepôt* of Southeast Asia. Historical background of the Malacca confirms that this patrimonial city-state was established out of nothing but the meeting point of international trades between East and West since the fifteenth century (Kratoska 2001, 97).

Geographically speaking, Melaka was not like any other typical Hindu-Buddhist town—*negara*—in Java or in other regions of Southeast Asia; as a city-state, Malacca did not have its own livestock production and existing local populations. The geographical location and setting of the port was not a large fertile land for growing rice and a well-established alliance of villages that surrounded its symbolically cosmic center. Geographically, Melaka is neither a fertile land for agriculture nor a strategic position for sea and land defense. Why was Malacca an important meeting point between East and West that stood out from the rivals such as Pasai and Aru in Sumatra? One crucial possibility was the Malacca's connection to and protection from the Chinese Empire under the Ming dynasties as earlier as 1405. Legendary Admiral Cheng Ho or Zheng He with his massive fleets visited Malacca on a regular basis. Until the Portuguese's conquest in 1511, Malacca was probably the first and an authentic example of an *entrepôt* state in Southeast Asia; it did not have its own commodity and its existing populations (Ricklefs 2001/2008, 23).

As *entrepôt* urban form and culture of Malacca were undoubtedly multicultural from the onset. Bandar Melaka was strategically in the middle of cohesion, negotiation, collaboration, and contestation of economic and political powers between East and West as well as between Chinese emperor from the East and Siam or Acehnese kingdom from the West. In order to minimize the danger from outside, the kingdom of Malacca embraced Islamic polity in an early sixteenth century. The transformation of the Malaccan kingdom from Hindu-Buddhist *kerajaan* to Islamic *sultanate* was likely to minimize the threat from the Acehnese kingdom. By giving the Chinese the concession of trade and stay in Malacca, the newly established sultanate was inevitably under the protection of the Ming emperor. Bandar Malacca was aptly protected by Chinese sea power. In return, the port authority, *syahbandar*, of Malacca was under the control of Chinese chief (Ooi 2009, 1286).

Multiethnic settlement belongs to Melaka's heritage since the fourteenth century. The traces of such settlement are still recognizable until today. *Kampung Serani* is the area where the Portuguese descendants live from the fifteenth century until today. The same case with *Kampung Keling* for Indian, *Kampung Cina* for Chinese, *Kampung Jawa* for Javanese, *Kampung Belanda* for the Dutch descendants. The traces of multiculturalism in Malacca is apparent until today; it is reflected by its diverse populations, a variety of culinary, mixture of urban architecture, an assortment of urban events, the array of domains and street names. The importance of multiethnic communities in the past had to do with the trades. Since modern times, the diversity of multiethnic populations has become the constructive elements of sociocultural interactions that have been shaping and developing the unique urban life in Malacca. Historical assets and heritage must have been functional in support of society today and future. Until recently, the hospitality industry is probably the main powerhouse for the sustainability of multiethnic liveability in Malacca.

Concerning its urban form, despite the city-state of Melaka was not located in inland geographical setting, the traces of the *mandala* system in the state polity were likely apparent with significant adjustments; the *Seri Nara Diraja*—the rule—was surrounded by four cardinal dignitaries consisting of *syahbandar* (port authority), *bendahara* (prime ministers), *temenggung* (chief of state security and public safety), and *laksamana* (navy commander). Under these four first rank state officials, there were eight dignitaries with titleholder of *Tunku* at the second rank and 16 titleholders of *Tan Sri* at the third and 32 titleholders of *Datuk* at the fourth level. The patrimonial state bureaucracy in pre-colonial Melaka was developed with Hindu-Buddhist statecraft system that posed the King at the apex of the mythic *Mahameru* cosmology (Walker 2009, 47).

Regarding its patrimonial tradition, the structure of state polity must have been symbolically incorporated in the urban form or in the site plan of the city center. Until recently, the traces of *mandala* juxtaposition remain inconclusive, due to the lack of archaeological finds and historical records of the pre-colonial city of Melaka (Manguin 2000, 410–1). Nevertheless, the waterfront location of Melaka is probably a challenging site for the installment of *mandala* symbolization. Nevertheless, the location of the former royal palace (*istana*) and port (*bandar*) was identifiable and recognizable as functional and symbolic positions. Of course, due to the waterfront setting, architectural adjustment for the layout became necessary. *Syahbandar* and *laksamana* belong to the water or the low position, while *temenggung*, *bendahara*, and *Seri Nara Diraja* stood on the high position in representing the land. In order to bring the downstream and upstream position functional, a vital communication road or path should be provided that must have been beyond the existing river. All this is recognizable in today site plan of the city. The dualistically spatial system based on upstream and downstream category is nothing new for orientation; most coastal communities in archipelago utilize such orientation for a practical reason. Even though the main street of the *bandar* Melaka could have been Jalan Laksamana and Jalan Hang Jebat, the symbolic axis of the high–low category is directly accessible to the river line.

The seat of the ruler of Melaka should have been on the hill area where the Dutch VOC established their stronghold. Logically, the Portuguese sea power under Admiral Alfonso Albuquerque conquered the Sultanate of Malacca in August 1511 (Ricklefs 2001/2008, 26) and then burnt down the palace of Sultan Mahmud. The Portuguese did not only build their stronghold on the site of Sultan Mahmud's palace but also their walled town called A Famosa on the whole area of the hill. The south riverbank area was mostly occupied by the elites and foreigners while the north side of the river was the native areas of *kampung*s. For safety and security reason, it is possible that the Sultanate palace and the *pekan* were at the hill of the south riverbank across the native settlements. Geographically speaking, the site of the palace was a strategic viewpoint to all parts of the city. The spatial division of pre-colonial Melaka is probably based on a dualistic category of despotic patrimonial state: the ruling class, *orang-besar* or *pembesar*, and the ruled or commoners, *rakyat* or *hamba-sahaya*.

As an urban center, Melaka should have been provided with a permanent market, *pekan*, and others places for public gathering. In today Malacca, such places are not easy to be found. However, the open area in front of the *Stadthuys* was a potential place for the pre-colonial market, *pekan*. The area was supported by a hub of streets from different directions; the site is strategically located for the meeting place of exchanges and trades between the natives and the foreigners. The importance of the market is not only for trades and exchanges, but also to enhance the sociocultural liveability of the city and the waterfront characteristics.

In the past, the liveability of Melaka is supported by the waterfront-oriented lifestyle by sea and river. During the sultanate period, the sustainability of *bandar* and its economic activities as an entrepot is indispensable of the Chinese protection; the Ming dynasty protects Malacca as well as works on bilateral cooperation with the Sultanate of Pasai in Aceh. Even though the Islamic city and state of Melaka is well known as a Southeast Asian cosmopolitan center in the sixteenth century, one barely finds on the whole picture of urban life during the Sultanate reigns from Chinese reports prior to the contacts with Europeans. Tarling describes Malacca city-state from Western sources based on notes of travelers and traders (Tarling 1999, 183–9).

Undoubtedly, political crises and power struggles took place during the period between 1390 and 1511. The city of Melaka was like any other city-state in Southeast Asia that was not free from ethnic tensions and political upheavals. All this was obviously demonstrated by the reigns of them (Ricklefs 2001/2008, 23), from Sultan Iskandar Shah (ca 1390–1424), Megat Iskandar Shah (1414–1424), Muhammad Shah (1424–1444), Hindu-Buddhist king Parameswara Dewa Shah (1445–1446), Sultan Muzaffar Shah (1446–1456), Manshur Shah (1456–1477) to Sultan Mahmud Shah (1477–1511). The other possible location of the Sultanate palace was the site on the hill where the Portuguese built their church: St Paul (Pieris 2009, 41).

As the capital city-state, Melaka was well known for their heritage of urbanism in terms of statecraft. The architectural heritage of Sultanate institutions of pre-colonial Melaka state must have been significant for an urban form of the *bandar* during the period of time between 1390 and 1511. The possible traces and signs of urban form during those periods were the local concepts of state and urban institutions such as *istana raja/sultan*, and various *jabatan jabatan* (offices): *bendahara, temenggung, syahbandar, laksamana*, various places: *bandar, pasar/pekan, kampung, medan*, and promenade.

The concentration of power in the period between 1390 and 1511 attracted the concentration of the Malay settlement at the confluence of

Melaka River. The Malay populations were needed for running day-to-day state bureaucracy and for strengthening the Malacca Royal army (*tentera darat*) and navy (*tentera laut*). Indeed, royal families and relatives who were mostly ethnically Malay occupied most state offices—*jabatan/jawatan Negara*. It is still unclear where the garrison areas were for Sultanate army and navy?

The Malay populations lived in various *kampung*s and settlement surrounding royal palace, *istana*. The elites lived on the hill whereas the commoners occupied the lowlands. As a state organization, Melaka's area would have been supported by three land use zone of institutions which consisted of areas for *pentadbiran* (state offices), *mahkamah* (royal courts and halls), and *kampung*s (residential). Almost all royal institutions disappeared after the fall of *Bandar* Melaka from the periods of Malay Sultanate to the Portuguese colonial rule in 1511. This fact gives us the possibility that the architecturally disappeared institutions were functionally replaced by similar ones with different names; they have probably similar functions or meanings. Albeit the sites, relics and artifacts of Royal Melaka institutions have been replaced, displaced, transformed, and modified by other forms, styles, and appearances, the idea and spirit of urbanism as the phenomenon of live–work–play in proximity and diversity for being *urbane* never die.

The multicultural reality of pre-colonial Melaka was represented by the existence of at least four Port Authorities—*syahbandar*—for major populations of the city: Indian Keling, Javanese, Malay, and Chinese ethnic groups. Each group lived in their specific territory; the traces of their origins are recognizable from the naming of place. Each port authority was provided with revenue section, *cukai*, which was directly under the line of command of the state treasury, *bendahara*.

Historically speaking, *syahbandar* or port authority was always the right hand of the Sultan for the economic and military value of the port. The number of *syahbandar* in a town or city showed the intensity of activities of the port; the more ships anchored at any one time was the more powerful was the port. Malacca in the fifteenth century had at least four port authorities with 2000 ships anchored at any one time (Ooi 2009, 1287). Indeed, the port for the sultanate was not only a place of trade. Rather, the seaport was mostly the starting point of political power and urban settlements in the Malay Peninsula because of its strategic position for exchange between foreign and native local commodities; the importance of port lies in its reality as the origin of civilization at the waterfront. The best example of this phenomenon was the port and *syahbandar* of pre-colonial Melaka; there were other places as well such as Kuala Terengganu, Pulau Pinang/George Town, and Johor Bahru—established in 1855 with old name Tanjung Puteri.

The development of vicinities at the Malacca Straits from port settlements to urban centers has been architecturally shown with the structure and form of *perbandaran*. Accordingly, the waterfront and its port is the core of urban life while settlements are built to support. *Bandar* is distinguished from the ordinary settlement of *kampung* because of its state institutions. Precisely formulated, *bandar* is because it has a principal port so that it is provided with the state institution of *syahbandar*. The live and sustainability of port and its waterfront as a gathering place characterize the existence of a *bandar*. In the Malay word, such a place is called *pekan*. The idea of urban settlement and society in the Malay language pertains to several concepts. The first refers to the concept of *pekan*, which is originally from the market, crowd, and exchange.

The second concept is *negara*, which is originally a Sanskrit word meaning state, authority, and capital city. The other concepts are *kota* and *bandar*. The origin of the word '*kota*' is probably derived from the Javanese word '*khuta*' or '*khita*' meaning center, fortified area for the seat of state, the source of power and order. However, it is unclear whether the Javanese words *khuta*, *khita*, and *kuta* are originally Sanskrit (Anderson et al. 2003, 94). In Javanese context, the word *khuta* or *khita* is irreplaceable in relation to and in association with the concept of *negara*, state or kingdom.

The characteristics of Malay urbanization in terms of *perbandaran* took its shape probably not earlier than the fourteenth century; it was the time when the state of Malacca was established. Were there other states before the Malacca? Available archaeological evidence in the Malay Peninsula has not yet confirmed us on this matter. However, the probability of the existence of states is open, which was likely under the control of Sumatran Sriwijaya or the Javanese Majapahit Kingdom, circa the thirteenth century. Historical evidence points out the name of Parameswara, a Prince of Sriwijaya, who fled from Sumatra to the peninsula. He established the state and port of Malacca. Under his reign, Malacca became an important Seagate of trade between East and West until 1511.

Studies on Malay urbanization are necessary to consider other urban centers in Southeast Asia. Since the early century, most ports and state capitals in the region had been involved in trades with China and India. Moreover, historically most states in the region were politically and economically related to each other. Trade of spices, silks, beads, gold, pearls, and indigo was an important activity in Southeast Asia since at the beginning of the century.

The pre-colonial intensive contacts and trades with international merchants were between the twelfth and fifteenth centuries. During those periods, Melaka at the East as well as Kuala Berang (Terengganu) at the West became the meeting points between Arabic-Persian-Indian traders and Chinese counterparts. It was probably the formative times for the urban characteristics of the Malay Peninsular *bandars*. Multicultural populations grew significantly during the periods mentioned above.

Despite the structure of indigenous settlement in Malacca, *kampung*, is loosely arranged without strict order, houses and places are connected to each other with a continuous walkway. As a neighborhood, a *kampung* is provided by its community with a mosque, Islamic school, *madrasah*, and open spaces. *Kampung* community is traditionally an egalitarian neighborhood. The leader of the *kampung* is usually a strong man in terms of knowledge on traditions and customs, as well as a religiously respectful person. In many cases, *kampung*'s chief called *penghulu* is a hereditary position. However, elders of *kampung* usually have a routine assembly to elect a new leader in case the old chief died. In their traditional village polity, aristocratic group and elders of ordinary people constitute the core of village assembly. In the coastal areas, the assembly is more egalitarian such as indigenous villages in Malacca.

The settlement of *kampung* builds and maintains their social cohesiveness on the voluntary foundation. In certain times, they work together voluntarily to maintain drainage, public water wells, and village safety and security; leadership of *penghulu* plays an important role in all this kind of work. During the pre-colonial eras, the Malay migrants from Java, Sulawesi, and parts of Sumatra such as Riau, Palembang, Jambi, Aceh, and Deli came to, lived, and worked in Malacca. They established and developed their settlement with a different title of village head such as *demang* for *Kampung* Jawa and *dato'* or *matowa* for *Kampung Bugis*. The Bugis village settlements in Malacca are mostly at the waterfront and become on-water habitation, *pakkaja*; in contrast to the agriculture based-village, *pallaung ruma*, the village of *pakkaja* is constructed as an architecturally interconnected house with continuous boardwalk. However, the *pakkaja* in many coastal places of the Peninsular Malay world has been established with Bugis traditional customary laws and tradition.

In the urban areas of Malacca, most *kampung* populations lived from working for the port, city, and the state as labors; some held official positions and did native commerce. Legends and local narratives tell people that state officers of Malacca Sultanate used to live with ordinary people in *kampung* settlement. Indeed, their residences are better in construction and building materials. Commonly, ordinary houses are built surrounding the prominent person's house. During the reigns of the Malay Sultanate, *kampung*s are only the main component of urbanism for the *bandar* Melaka, but also are the most important part of the state that upholds and sustain Royal traditions and rituals in Islamic ways. However, the Muslims in *kampung* used to live with other ethnic groups, such as Chinese, Gujarati, Arab, and Persian populations.

Like the indigenous Malay populations, other ethnic groups build and establish their communities with their own customs and traditions. However, religious facilities are not always the center of their settlement. The Chinese populations, for example, live and work as traders in their shophouses. The center of activities for the Chinese communities is the commercial street where their shophouses are. The meeting area for all ethnic groups is located in the market, *pekan*. Traditionally, the market of Malacca Sultanate is actually for not only international trade but also for the fish and general market.

Traditionally, *pekan* is an occasion and gatherings of people on the permanent area; the people come from various villages for the event that happens in every other five-day period. *Pekan* means literally the period of five days in the moon-based calendar system. During the event of *pekan*, people do not only exchange foods and goods, but also ideas, works, and performances. The existence of *pekan* is the heart of the Malay community where sociocultural and economic activities of *Bandar* take place. In a matter of fact, the origin of *pekan* in the Malay world is probably rooted in the Hindu Malay state tradition. The *pekan* is originally the farm market of various villages that take place in the capital of the state, *negara*.

During the reigns of Islamic Sultanate, the phenomenon of *pekan* has been elaborated and enhanced with the concept of Persian *bazaar*. In the Malay word, the *bazaar* is adopted and customized with the word *pasar*. The development of the market from *pekan* to *pasar* signifies the importance of this spot as an international exchange place. The establishment of the market from *pekan* to *pasar* was an integrated part of the development of *bandar*. *Pasar* is distinguished from *pekan* because of its permanent service during the daytime for all people; it is not an event every other five-day period. As a matter of fact, each ethnic communities in pre-colonial Malacca had their own *pasar* and port-authority, *syahbandar*. Historical Chinese records from the Ming dynasties show the pre-colonial Malacca having at least five port authorities.

Multiculturalism in Malacca 1511–1800

The presence of Chinese merchants in Melaka was not only by the nature of trade but also by bilateral agreement between the Malay Sultanate and the Chinese Emperor under Ming dynasties. Accordingly, Chinese populations were free of tax to live and work in the city. A most first wave of Chinese migration to Malacca from circa the sixteenth century was originally from various parts of China. The second wave of Chinese populations was in the nineteenth century mostly from Guangdong and Fujian who speak Hakka, from Quanzhou and Amoy who speak Hokkien, from Guangdong and Guangxi who speak Cantonese, and a small number of people from Chaoshan who speak Teochew. Unlike the later migrations, the first Chinese migrants consist of various social milieus and occupations. The later migrants were mostly jobseekers for working in rubber plantations and traders.

One important contribution of the Chinese community in Melaka to multicultural urbanism was the presence of their shophouses. Of course, as an architectural unit of urban form, the Chinese shophouse had evolved through times. The form is remarkably an architectural response to urban proximity with a mixed-use system of shop and house, as well as with a compact and strong design for street edge and urban block. Regarding its low-rise structure, the shophouse architecture potentially creates and enhances a streetscape of urban form with human scale and repetitive rhythm. All this is optimally successful with the support of building codes and urban ordinances for consistency in terms of the use of style, form, structure, and building material.

Historically, the form of a shophouse in Melaka was indispensably functional in terms of the unity of shops and houses. In doing so, shophouses are home and workplace that support the liveability of urbanism at street-level activities. The reason for this fact is quite simple for safe, effective, and efficient urban retail and services. How did the early version of shophouses in Melaka look like? The question leads us into the possible adaptation of the Malay house form with their local materials and construction. The Malay and the Chinese residential architecture in the sixteenth century likely shared something in common concerning the wooden structure and gable roof form. However, the Chinese counterpart did not likely build their building raised on stilts, as found in the Malay house form. Instead, the Chinese shophouse designed their buildings with a ground floor higher than the land surface (Fig. 9.1).

Architecturally speaking, the presence of Chinese group in the fifteenth century on likely contributed the formative urban block and urban character of Melaka city with low-rise high-density settlement. Even though the use of building materials in the fifteenth century was predominantly wooden and bamboo construction, an early urban block of Melaka could have been defined and developed by the sites occupied by the Chinese row shophouses. In contrast to the Malay *kampung*s with loose units of living house, the Chinese settlements in Melaka built a compact row block of shophouses; prior to the arrival of European people, they were not built with a brick-layered bearing wall structure. The brick-layered building could have been the influence of Dutch brick town house between the seventeenth and eighteenth centuries.

Moreover, the Chinese and Indian community in Melaka enhanced the city with unique Chinese Buddhist temples and shrines such as Cheng Hoon Teng temple and the Kampung Keling Mosque. Like the Chinese, the pre-colonial Indian, Persian, and Arab populations in Melaka worked

Fig. 9.1 Multicultural relics on the streetscape of Malacca (Photos by author)

and lived from the port economic activities of four Melaka *syahbandar*s. They were mostly traders and businesspersons who traveled for trading for spices and silks. Their relation to Malay Melaka communities was not simply for business but also involved in a sociocultural relationship with the native. They introduced Islam and Middle Eastern culture to the native people in the coastal areas of the archipelago that included literature and poetry.

The Arabs and Persians introduced and enriched the native urban settlement with new concepts such as *madrasah*, *mukim*, and *pasar*. Even though *pekan* and *pasar* are identical, the last is derived from the concept of the Persian *bazaar*, meaning a permanent place for exchanges. *Pasar* is slightly different from *pekan* because it operates on a daily basis, not on weekly basis. Architecturally speaking, the presence of Muslim in Melaka was represented by a variety of mosque architecture.

There must have been several Islamic buildings during the reigns of Malacca sultanates. However, there were few artifacts surviving until today. Local narratives told us that there were mosques in almost Malay *kampung*s in Melaka such as the Kampung Hulu Mosque. Another important Islamic heritage in Malacca is the Kampung Keling mosque at Jalan Tukang Emas. Although the original was built with wooden construction on the same location, there is no further information when was the original mosque built? The today building was not the replacement of the original due to its demolishing building materials; the new mosque was established on the location in 1748 (Tadjuddin et al. 2005, 71).

Despite its perishable buildings, the embracement of Islam by most Malay populations contributed to the building of the sense of community in a territorial area of the *kampung*. Islamic rituals and institutions such as *surau, mosque,* and *madrasah* strengthened and enhanced the existing customs and traditions with more intensive gatherings and collaborations that potentially built a strong community in pre-colonial Melaka. All this was demonstrated by the fact that each *kampung* had its own mosque. In other words, the *kampung* settlements in pre-colonial Melaka were parts of municipality or city-state of Malacca. In Islamic statecraft, it must have been the law to determine people living and working in the state with specific political status either as citizen, *mukim* or foreigner, *warga asing.* In other words, the sense of urbanism in *bandar* Melaka was not only established with its architectural urban fabric, its monuments, and places, but also with the political status of its populations. All this was useful for the taxation and functional qualification for state occupations (Fig. 9.2).

Melaka under the Malay sultanates was probably not that perfect as what is told in local narratives such as *Sejarah Melayu.* However, pre-colonial Melaka city-state in the period between the fifteenth and seventeenth century was not without competing with other city-states such as Pasai, Palembang, Banten, Tuban, and Ayutthaya. Maintaining a good relationship with neighboring city-states was likely the key to the sustainability of *bandar* in terms of economy and society. Interracial marriage was a common solution in dealing with multiculturalism in the pre-colonial city-states in Southeast Asia. Pasai and Melaka were not only in a peaceful relationship based on a political and military treaty but were bound with intermarriage between both sultanate families and relatives. The problem

Fig. 9.2 Kampung Keling Mosque, Malacca (Photo by author)

urbanism is, in a matter of fact, not that simple. The diversity of ethnicity challenged urbanism in pre-colonial Malacca with various issues concerning hegemony and social justice. The despotic patrimonial state like the Malay Malacca in the fifteenth and seventeenth centuries was probably not a sustainable polity system for urbanism today. However, its sociocultural heritage is one important asset for urban local characteristics and features in Southeast Asia.

In terms of the urban institution, the Arab and the Chinese populations had a significant contribution to the establishment of the education system in the pre-colonial Malacca. Newbold (Thomas 1993, 170–3) describes in his report that there were schools for Malay and Chinese populations. The Islamic schools, *madrasah*, were mostly built around the indigenous *kampung*, while the eight Chinese schools were located in different places. Both school systems had their own emphasis. The Malay *madrasah* was more to teach the Islamic teaching with training to write and recite the

Al Quran. On the other hand, the Chinese schools trained their children on Confucian ethical philosophy and ancient Chinese literature, history, graphic art, and poetry. During the colonial period, the Dutch and the British rule established their schools as well.

The first free school in Malacca was established under the patronage of Governor Thyssen in 1819 (Evangelical Magazine and Missionary Chronicles 1823, 180). The school was open for European, Chinese, Indian, and Malay children. The schools were the first multicultural urban institution that attempts to integrate various languages and cultures into three areas of training: arithmetic, culture, and literature. There was no single official language in this school; Dutch, Chinese, and Malay were practised on a daily basis. However, for the first time, the school introduced the Latin alphabet for reading and writing. Based on this school, the Christian societies developed and established the Anglo-Chinese College in Malacca.

European Colonial Heritage in Malacca

Malacca is regarded as one of the historic urban settlements with highly concentrated colonial heritage in the Malay Peninsula. The architectural heritage is obviously dominated by various Dutch shophouses, homestays, offices, cemetery/tombs, places, churches, and warehouses. The contribution of colonial architecture in this city is apparent in shaping urban form with a strong and permanent structure of built forms and spaces. The Portuguese, Dutch, and British empires contributed to the formation of the urban structure of Malacca. The arrival of Europeans develops local building tradition from carpentry to masonry and bricklayer construction. The enrichment of architecture in Malacca comes into being with the hybrid of highly sophisticated carpentry from Chinese tradition and masonry construction of the Europeans. The hybrid of the European and Chinese presence in Malacca is apparently shown in the use of metalwork for door and window hardware and accessories, such as safe lock system, hinge, handle, and trellis for shophouses and warehouses.

During the sixteenth century, the Portuguese and Spanish Empire expanded their sea supremacy to control sea trades all over the globe. The successful expedition of Vasco de Gama to South Africa and Columbus to

Fig. 9.3 Portuguese, Dutch, and Malam Iconic elements in Malacca Malaysia (Photo by author)

America opened further explorations to the regions of spices in Southeast Asia. The Portuguese sea power came to Southeast Asia to take over the trade control of spices from Far Eastern archipelago to Europe. The straits of Malacca have been always a strategic route of trade between India and China. The Portuguese sea power came to Malacca and then took control of the international trades from 1511 to 1641 (Fig. 9.3).

The Portuguese sea power did not develop the site of former Malacca Sultanate with the urban settlement, but a simple port town with fortification for their warehouses and military barracks. Indeed, Catholic missionary was part of Portuguese explorers coming to the Far Eastern countries including Malacca. The missionary led by Francis Xavier served Portuguese soldiers and officers in the foreign land between 1541 and 1549 as well as introduced the native populations to Christianity; they built the Catholic Church—*igreja*—St. Paul on the hill of Malacca. Important landmarks under Portuguese colonial influence include remarkable buildings of porta, fort), *igreja* (Church) *praca* (place/square), *marcado* (market) and *campo* (field/public open space), *Padang Portugis*. Of course, they did not build exclusively *palacio* (palace/townhall) and *praca* (special bullring) in Malacca. Today, the Dutch, British, and modern Malayan influence might have overlapped and imposed such places with different names but similar function. Today, *padang* Portugis in Malacca is still in existence as an open public place but the populations surrounding the campo have changed. The Portuguese Eurasians live and work today in various places and urban areas of Malacca.

The most important landmark of Portuguese heritage is the A Famosa meaning famous; it is a Portuguese fortress marking the success of Admiral Alfonso de Albuquerque in conquering the Malaccan Sultanate in 1511. Together with Macau, and Goa, the fortress of A Famosa was envisaged by Albuquerque as an important and strategic post of Portuguese sea power controlling the spice and silk route from the Far East to the West. The function of the fortress was a military garrison with warehouses and food storages. Architecturally speaking, the fortress had an exceptional bricklayer structure. The remnants of the bastion show high-quality bricks and Vauban-design style of the citadel with some watchtowers and four corners.

The presence of Portuguese populations in Malacca has been contributing to urban life since the seventeenth century, such as street festivals, folklore, and gastronomies. The Portuguese descendants in Malacca mostly have been living in *Kampung Serani*, the native Malayan name for their Christian village. In the artistic field, Portuguese explorers introduced the indigenous people in coastal areas to the musical performance of *fado* with guitar and a vocal singer. To a certain extent, the melancholic, mournful, and homesick tone of *fado* has influence in the Malay musical composition and performance until today. Indeed, Arabic and Persian musical traditions have significantly been contributing to the musical appreciation and development in the Malay world as well.

The Dutch colonial power defeated Portuguese control over Malacca in 1641. Like the Portuguese empire, the Dutch VOC (the united far eastern companies) established their trade post in Malacca after they destroyed parts of the Famosa. The Dutch power built their trade post on the West of former Portuguese stronghold. The Dutch ruled and colonialized Malacca

Fig. 9.4 The ground plan of A Famosa, Portuguese Fortress in Malacca (Photo by author)

for their trade post and port on the route to the Indian Ocean until the beginning of the nineteenth century. Gradually, the Dutch populations increased and developed Malacca as a colonial town with the architecture of brick shophouses (Fig. 9.4).

The presence of the Dutch populations in Malacca reached its peak circa in the period between 1780 and 1830 that led them to integrate their bastion with the Chinese settlement and the indigenous *kampungs* at the west bank of the river Melaka (Clearly and Kim 2000, 140). More and more Dutch shophouses were built on urban blocks, which were architecturally defined by the network of streets. One famous street of the Dutch heritage is today Jonker and Herren Street. Unlike the Portuguese sailors and workers, the Dutch populations did not mingle with the indigenous people or Chinese populations. The Dutch established their own pubs and clubs; the famous places for Dutch socialites are now known as the taverns of Heeren Street.

The contribution of the Dutch presence in Malacca is demonstrated with various landmarks, structures, and places that include the Red Stadthuys, the Red square, several shophouses, bridges, windmills, offices, and houses. Like the Portuguese, the Dutch presence introduced Malacca to bricklayer building construction for various building types. Steel and wooden frame structures were brought by the Dutch to enrich and elaborated on the previous indigenous and Chinese building tradition.

In the context of urban settlement, the Dutch presence developed urban structure and form with permanent buildings and roads with drainage and sanitary system. The idea of compact land use and urban form was, in a matter of fact, without planning and design, but simply pragmatic decisions and actions based on the Dutch experience in their homeland. Under the Dutch administration, intensive interactions between Malacca and Batavia, as well as other ports of archipelago such as Belawan, Makasar, Semarang, and Surabaya, developed cultural acculturation of European, Chinese, Indian, and Malay culture. The outcome of this acculturation of interracial and multicultural interaction was the phenomenon of *baba-nyonya* urban community.

The Napoleon War in Europe in the beginning of the nineteenth century brought about the transition of colonial power from the Dutch to the British rule. Under the British control, Malacca was less regarded as a trade port and town on the Malacca Straits. The British Empire preferred to do their business in Georgetown of Penang. The British rule did not want Malacca to be in rivalry to Georgetown. Instead of maintaining and developing Malacca, the Governor General planned to demolish the Malacca fort and to move 15,000 people from this town to Penang in 1807. However, Sir Stamford Raffles in his capacity as the Lieutenant Governor of the British Empire in the Far East managed to rescind the demolition of the Fort and depopulation of the town in 1818.

During the British colonial period of 1825 and until the early of the twentieth century, the urban development in Malacca was more in the hand of the private sector. Even though many Dutch populations were expelled and relocated from Malacca to other towns and cities in Dutch East Indies, some of them stayed and worked in town running their businesses (De Witt 2007, xix). Other European, Arab, Persian, Chinese, Indian, and indigenous merchants and traders took over the Dutch

Fig. 9.5 The heritage of Dutch Colonial shophouses in Malacca (Photo by author)

community's role; they played a very important role in the economic activities in the town. Interracial marriages were evident in Malacca of the eighteenth and nineteenth centuries; the Malay word for the descendants of interracial couples is *peranakan*. Since the Dutch rule left Malacca, Chinese, Malay, and Indian people built, lived, and worked in the town with specific professions that supported the urbanized populations of trade and port town. Grouping and concentration of similar businesses took place along the streets of the town, such as iron and goldsmith, restaurants, home stays, handicrafts, herbal medicines, clothes and textiles, goods, and foods (Fig. 9.5).

Malacca with its predominantly established Portuguese and Dutch architectural heritage had made the British colonial rule feeling uncomfortable to establish their center in the Malay world in this city. As a matter of fact, the British legacy in Malacca is architecturally unremarkable. Nevertheless, in 1904, the British Empire had tried to fasten their foot in Malacca by installing the fountain in front of the Red Stadthuys; the fountain was dedicated to the Queen of Victoria. Probably, Raffles had already realized this futile attempt and moved to Singapore to establish a British colonial center there. For him, building the new world on the new land was probably more dignified rather than that on the old one by destructing and demolishing the precedence.

In general, the contribution of colonial regimes in Malacca was to introduce the Malay world with the Western civilization. Christian missionaries from various groups played an important role in introducing the Western literate culture to the indigenous and Chinese populations. The Western missionaries have established Churches and Christian societies in Malacca since the seventeenth century; their churches become important landmarks of the Malacca urban architecture. Since the Dutch took over Malacca from the Portuguese, several Catholic churches were abandoned or destroyed (Hussin 2007, 277). The centenary victorious presence of the Dutch rule over the Portuguese was celebrated with the erection the Evangelical Church on the St. Paul Hill. Several Catholic churches in the urban area were transformed into the Reformist Churches. However, existing Portuguese populations, especially in *Kampung Serani*, still maintained their church: the Immaculate Conception built in 1811, their *campo* and *mercado*.

Moreover, financial institutions such as banks and insurances were modern agencies of urbanism that provided Malacca with internationally acknowledged *entrepot* of the globe since the eighteenth century. The decline of this status began the British Empire decided to choose Penang as their Far-Eastern *entrepot*. Later on after the opening of the Suez Canal in 1869, the British power moved their trade center in the Malay world from Penang to Singapore. Prior to 1920, there were no towns and cities in the British colonized Malay Peninsula, which was administered with regulatory development control. The first concentration of settlement established a planning and development section was Kuala Lumpur in 1921. The purpose of this installment was to manage the development of urban infrastructure. The urgency of planning system under the British colonial administration was in response to the risk of flood and unhealthy sanitary system.

In terms of housing planning and design, the British colonial rule introduced terrace houses in a double row block system divided by a back lane. One block system usually consists of 10–16 houses/acre or 4050 square meters; the common density of the residential area is 25–40 units/hectare. The British rule introduced and implemented the layout and terrace house design with back lanes in their colonial towns of Georgetown and Kuala Kubu Bharu in circa 1930s. Since then, urban structure of Malaysian towns and cities today is characterized by the British colonial legacy of urban blocks consisting of two-story residential units as well as shophouses, which are linearly arranged with a back lane.

Post-colonial Urbanized Area: *Perbandaran*

Under the notion of *perbandaran*, the Malay people understand a politically organized settlement with at least the following traditional institutions: *istana* for royal palace, *pasar/pekan* for permanent market, *masjid raya* for great mosque, *mahkamah* for court of justice, *majlis perbandaran* for townhall, *dewan raya* for great meeting hall, *kubur* for cemetery, *mukim* for residential subdivision, and *kampung/taman* for village or neighborhood. Traditionally, most *bandars* were provided with a port authority of *syahbandar*. Today, *bandar* is more about the general idea of an urban settlement. The importance of this institution lies in its direct relation to the state.

Traditionally, the Sultan is a traditionally spiritual lord with a concentration of power at his hand. During the Colonial rule and Post-Colonial government, the power of Sultanate in each state is more associated with state ceremonies and rituals, rather than with direct and day-to-day state business. In other words, Sultanate power in terms of the planning system is more politically symbolical, rather politically practical. As a symbolic center, the Malay Sultan stands at the apex of the spiritual pyramidal structure of power. The second level of the structure is the four dignitaries who are in charge for specific duties and functions in the state bureaucracy. The two subordinates of each of the four dignitaries occupy the third layer of power. As a whole, the idea of power in the Malay state and city reflects a hierarchical and feudal structure of the decision-making process. Individual preferences and personal interests of the Sultan play a dominant role in the life and characters of the city and state.

Since the establishment of the Federation of Malaysian States and Nation in 1957, citizenship of urban areas has been to comply with the Malaysian citizenship based on blood tie. Malaysian citizenship regulates all residents of age 12 and over to possess Malaysian identity card. The ethnicity and religious affiliation are shown on the identity card. The definition of Malay individual is enshrined in the Constitution as a person who professed Islamic religion, habitually speaks the Malay language, and conforms to Malay tradition and custom. The Malay populations and other indigenous people, *orang asli*, share their right to be called *bumiputera*—son/daughter of soil. The concept qualifies an individual as Malay or *orang asli*—indigenous person—that deserves special rights through the constitution (Hestflatt 2003, 69).

Malaysia has three major ethnic groups: Malay, Chinese, and Indian populations. Regardless of their race, all citizens are due to pay taxes to the Federation of States. Foreigners have to report their income tax but will receive the return. The privilege of urban citizens in this country includes the right to vote for public election, ownership of property, benefits for education, health insurance, housing, utilities, and loans for entrepreneurship. However, racial category determines the grade of some privileges from the Federation of States; until now, the Malay populations enjoy such advantages of the government's loan for setting up businesses and homeownership. The Malaysian government regulates the benefits for the Malay, Chinese, and Indian populations differently.

Malaysian urban society is racially segregated in terms of beliefs and rituals. Ethnic and cultural identities in this country are associated with their religious affiliation (Saw and Kesavapany 2006, 18). Muslim Malays are prohibited by Islamic laws to convert to another belief or to renounce Islam. The three major ethnic groups have the right to exercise their customs and traditions in public domains and schools. Each group has their own school system with a strong affiliation with Islam for the Malays, Buddhist and Animism for the Chinese, and Hindu for the Indians. The celebrations of religious events and rituals characterize and feature urban life with unique occasions such as *Raya Ied*, Chinese New Year, and

Deevawali. Indeed, the dominant Islamic populations play an important role in practicing codes of conducts and manners in public domains.

Demographically, post-colonial urbanization took its tall after 1970 in Malaysia (Yaakob et al. 2000). The period between 1970 and 1980 was the time when the economic recovery of the country came into a more stable condition. Prior to 1970 census, the urban settlement was defined as an area of inhabitation with populations of 10,000 or more; since those time mobility of populations from rural to urban areas became growing. While the Federal Land Development Agency, FELDA, was actively handling the urbanization and resettlement of rural people by providing access to the land, the government held the trend with an urban development policy for the development of new urban areas. The drastic transformations of peninsular Malaysian landscape happened between 1970s and 2010s. Mergers, amalgamations, and consolidations of urban areas took place during these years.

Features of Modern Urbanism in Malacca

The 1970s was the year when the landscape of urban settlements in the Malay Peninsula marked its milestone of economic and political stability (Saleh and Meyanathan 1993). Since then, the urban landscape and its settlements began to grow in response to demographical and economic development. Prior to 1970s, few scholars were interested in the urban studies of the region. Studies by several scholars (Manguin 2000, 409–30; Evers and Korff 2000; Bishop et al. 2003; Suhaimi 1993, 63–77; Wheatley 1983) demonstrate that urbanism in Southeast Asia not only is interesting in terms of its drastic transformations and changes but also shows us that there are many things to be investigated and studied carefully. One of them is a study on the origin and evolution of the concept and phenomenon of waterfront multicultural vicinity: *bandar*.

Until now, such a study remains less explored if it does not fall into oblivion at all. The unique characteristics of *bandar* as urban settlement and its society have not been profoundly described and studied as a poten-tial contribution to a theory of urbanism with multicultural background and history. The issues of urbanism in the region were and are mostly about the concentration of settlement in the crowd of multiethnic diversity and hub of international trades. The waterfront vicinities between the Southeast Asia peninsula and Sumatera Island have been an international prominence of trades between Far-Eastern and Near-Eastern countries since the fifth century (Sheriff 2010, 171–90; Leifer 1978, 6).

Dining out in outdoor and in public domains is the most favorite activity in leisure time for most populations in the Malay towns and cities; they do together with the whole family member in the mild and clear evening. Various restaurants offer local foods and drinks with their own special menu. Interestingly, foods and drinks for the Malay populations have to be stately certified *halal* and free from pork and lard. With such certification, the Malay could go into Indian and Chinese restaurants or cafes. To a certain extent, religious barriers and cultural differences do not make interracial interactions among them easier. Even though the grouping of informal gathering in public domains is based on racial identity, they share something in common for peace and tolerance.

The reality of ethnic relations and multicultural society in Malacca is likely not without challenges. The Eurasian Portuguese in many parts of Malacca area have to face the crisis of status and identity after the establishment of the Federation of Malaysia. Even though their existence as a community in the Malay world has been since the Dutch colonial rule, the integration of their ethnic status and identity into the Malay political party and society is still the work in progress. Even today, *Kampung Serani* remains only the name of the site; the residents of the Eurasian village were not around in the area anymore. The relationship between ethnic populations and its cultural heritage in today Malaysia is in a crucial situation that must have been in the context of power relation with the predominant Islamic populations.

The Malay populations mostly occupy public sectors, while Chinese and Indians mostly run businesses and other private sectors. Migrant workers fill the gap of occupations, especially for labor in manufactures, utilities, commercial services, and domestic jobs. In urban development and construction, foreign workers fill most laborious positions; employment agencies arrange and manage rent flats and accommodations for such workers. As a whole, the image of the Malay towns and cities is multicultural and vibrant during the daily bright leisure evening.

Besides its diverse ethnicity, Islamic gender relation comes into play in public domains. Women and men suppose not do handshake when they meet. Malay, Indian, and Chinese populations share their value for not hugging each other when they encounter each other in public places. Interpersonal interactions in public areas are strictly polite and quite formal. The common language for interracial communication is mostly Malay language. However, they know and speak the English language to each other as well.

Concerning their existence as a nation, the three major ethnic groups have been trying their best to live and grow together, given that their schools are racially segregated from early childhood on. One important platform for multiethnic interaction for nation-building is at the higher education level such as internationally rank universities, state institutions, and public sectors. The other important public domain where all ethnic groups mingle together is the town market—*pasar awam*. Naturally and informally, people interact with each other with the exchange of words and smiles. Unlike malls, the market is not visited by all family members; most visitors are from middle and working class. Nevertheless, the real and potential place and the process of nation-building are probably the marketplace (Fig. 9.6).

Urbanity is not perfect without taking manners and courtesies of its urban citizens in public domains. Significant efforts and developments for the pedestrian-friendly environment have been made in several towns and cities such as Melaka's Hang Tuah Mall and Kuala Lumpur's Bukit Bintang strip and Central Market passage. Hang Tuah Mall in Malacca is an innovative pedestrian-friendly environment in town. The sidewalk has been extended and improved with amenities and landscape vegetation. The problem is that the outdoor mall is not conceptually designed as a connecting course of places. Hang Tuah mall remains mostly for destined visitors but not for passerby people; it does not have two strong generators of movement between Dewan Hang Tuah and the Plaza of Boon Seng that activate the street livability of Hang Tuah sidewalk. As a destination, the ample sidewalk at the Hang Tuah Street is appealing—many food stalls and outdoor restaurants and cafes. The 1500-m-long sidewalk was initiated by the Prime Minister Mahathir Muhammad, constructed on January 2, 2002 and completed on September 15, 2002. However, the

Fig. 9.6 Malacca Night Market 2012 (Photo by author)

integration of mass-rapid-transit and pedestrian path network is still the work in progress. Surprisingly, such integration seems not apparent for the government's new towns development such as in Putrajaya, Syah Alam, and Iskandar. Private car-oriented lifestyle dominates mostly public spaces and commercial areas (Fig. 9.7).

As anywhere on the globe, malls are the generators of traffic flow. Unlike in North America, malls in Malaysia are not only built in suburban areas but also become an integrated part of the urban centers, such as Petronas Twin Tower and KLCC in Kuala Lumpur, City Square in Johor Bahru, and Grand Market in Kota Bahru of Kelantan. In Malacca city, the important mall is located on the Ground of Heroes, *Dataran Pahlawan,* where the First Prime Minister Tunku Abdul Rahman proclaimed the independence of Malay nation-state on August 31, 1957. Now, *Dataran Pahlawan* has lost its memorial character as a historical site. As a whole, old town center of Malacca has not yet been integrated as an urban design system that provides the inhabitants, workers, business owners, visitors, and maintenance people a pedestrian-friendly environment and network with the

Fig. 9.7 Hang Tuah Sidewalk in Malacca (Photo by author)

safe and healthy atmosphere. Pooling parking areas and restricted car zones have not yet implemented that are connected with a convenient and safe network of bike tracts and pedestrian paths.

In order to explore the significance of historical artifacts and places, it is necessary to design several scenarios of urban historical experience. In doing so, visitors are able to have options of movement and visit for visual, lively and culinary experience within a certain category according to certain historical events of Malacca or precinct or location uniqueness. This includes new innovative development in old town area, for example, a promenade along the riverfront area for multicultural cafes, restaurants, and homestays (Fig. 9.8).

Providing specific area as a unique place of destination is a workable option in Malacca because of its historic area, scale and intimate spatial

Fig. 9.8 Waterfront area of Malacca (Photos by author)

and building proportion for pedestrian environment. In other words, the grouping of places in the inner old town area into historical clusters will make urban design treatment and improvement in Malacca easier and more architecturally focused. Safe, convenient, and continues network pedestrian area is still problematic in Malacca old town area because of ineffective management of restricted car area and insufficient service of public transit. As a whole, Malacca inner town area needs a building regulatory framework and principle that does not allow and admit new developments that are not in alignment and harmonious relation to the existing historic buildings in terms of scale, proportion, dimension, and function. Morphologically speaking, the inner town area entails small and intimate scale of buildings and places on parcel site. Superblock development with large as well as high-rise building seems unacceptable in the historic district of Malacca.

Concluding Notes

The idea of *bandar* in pre-colonial Malay world was a characteristic urban settlement at the waterfront setting. The concept was established and developed with the influence of Indian, Persian, and Arab civilization. In the beginning, the concept of *bandar* was about a city-state settlement with a port. The most important structure of *bandar* lies most likely in the existence of a well-established state organization, *kerajaan* that consisted of *raja, bendahara, temenggung, laksamana, and syahbandar.* The port authority was a strategic state institution that managed and secured trades and exchanges in a sustainable way because of its solid coordination with the revenue agency, *cukai.* In the case of Melaka, the sustainability of the *entrepot* status depended obviously on the protection of the superpower of the Chinese Empire under Ming dynasties. The fall of Melaka to the Portuguese emperor showed us how prone the *entrepot* was; it was not because of only the lack of hinterland support for human and natural resources, but also that of geopolitical dependency on international power relations. Nevertheless, *bandar* remains a unique phenomenon of urbanism in Southeast Asia regarding its cultural acculturation for various traditions, customs, and languages that potentially build and develop innovative ideas and things because of its diversity and competitive challenges.

The sustainability of urbanism in the Malay world is economically and culturally supported by a long-standing multiethnic collaboration and mutuality. Even though each ethnic group has its own way to preserve and sustain their customs and traditions, tolerance among them and the necessity for living in peace with each other prevail and overcome the potential conflicts and clashes among them. Despite under different regimes, Malacca had achieved the Golden Age of global and multicultural urbanism since the sixteenth century until the early of the nineteenth century. The key successful pre-colonial Malacca lies in the Sultanate policy for providing an autonomous domain for each community; they have the right to exercise their own customs and traditions. The interactions of diverse communities take place in the markets and public places. Since each ethnic community has its own specific leverage in terms of skills, products, and services, the diverse communities create an interdependency and cooperation that works within the web of economic exchange. Since then, Malacca has been always potential for a global and multicultural city, but the most challenging issue for this in the sustainable future is likely more about the political will for urban quality of living with heritage. Accordingly, Malacca needs a specific and long-term development program that enables its municipality to establish and grow its historical and geographical potentialities as an environmentally walkable city with less private cars and well served by the green public transit system.

References

Anderson, Benedict, James Siegel, and Audrey Kahin. 2003. *Southeast Asia Three Generations: Essays Presented to Benedict Anderson*. Ithaca, NY: SEAP Publications.

Bishop, Ryan, John Philips, and Wei Yeo Wei. 2003. *Postcolonial Urbanism: Southeast Asian Cities and Global Processes*. London and New York: Routledge.

Chambert-Loir, Henri. 2005. "Sulalat al-Salatin as Political Myth." *Indonesia* 79: 131–60.

Clearly, Mark, and Chuan Goh Kim. 2000. *Environment and Development in the Straits of Malacca*. London: Routledge.

De Witt, Dennis. 2007. *History of the Dutch in Malaysia*. Kuala Lumpur: Nutmeg Publishing.

Evangelical Magazine and Missionary Chronicles. 1823. *Memoir of the Late Rev. Milne D.D.* London: Francis Westley.

Evers, Hans-Dieter, and Rudiger Korff. 2000. *Southeast Asian Urbanism: The Meaning and Power of Social Space*. Muenster: LIT.

Hestflatt, Kristin. 2003. "The House as a Symbol of Identity Among the Baba Melaka." In *The House in Southeast Asia: A Changing Social, Economic, and Political Domain*, edited by Stephen Sparkes and Signe Howell, 67–82. London: Routledge.

Hussin, Nordin. 2007. *Trade and Society in the Straits of Melaka: Dutch Melaka and English*. Copenhagen: NIAS.

Kratoska, Paul. 2001. *South East Asia, Colonial History: Imperialism Before 1800*. London: Taylor & Francis.

Leifer, Michael. 1978. *Malacca, Singapore, and Indonesia*. Dordrecht: Brill.

Manguin, Piere-Yves. 2000. "City States and City State Culture in the Pre 15 Century Southeast Asia." In *A Comparative Study of Thirty City-State Cultures: An Investigation*, edited by Mogens Herman Hansen, vol. 21, 409–30. Copenhagen: Kongelige Danske Videnskabernes Selskab.

Ooi, Keat Gin. 2009. *Historical Dictionary of Malaysia*. Lanham: Scarecrow.

Pieris, Amona. 2009. *Hidden Hands and Divided Landscapes: A Penal History of Singapore's Plural Society*. Honolulu, HI: University of Hawaii Press.

Ricklefs, Merle Calvin. 2001/2008. *A History of Modern Indonesia Since c. 1200*. Stanford: Stanford University Press.

Saleh, Ismail Mohammad, and Saha Devan Meyanathan. 1993. *The Lesson of East Asia, Malaysia: Growth, Equity, and Structural Transformation*. Washington, DC: The World Bank.

Sandhu, Kernial Singh, and Paul Wheatley. 1983. *Melaka: The Transformation of a Malay Capital c. 1400–1980*. Oxford: Oxford University Press.

Saw, Swee-Hock, and K. Kesavapany. 2006. *Malaysia, Recent Trends and Challenges*. Singapore: Institute of Southeast Asian Studies.

Sheriff, Abdul. 2010. *Dhow Cultures and Indian Ocean: Cosmopolitan, Commerce, and Islam*. New York: Columbia University Press.

Suhaimi, N. H. 1993. "Pre-modern Cities in the Malay Penninsula and Sumatra." *Jurnal Arkeologi Malaysia* 6: 66–77.

Tadjuddin, Mohammad Rasdi, Komarudin Mohammad Ali, Syed Iskandar Syed Arifin, Ra'alah Mohammad, and Gurupiah Mursib. 2005. *The Architectural Heritage of the Malaysian World: Traditional Houses*. Skudai: UTM Press.

Tarling, Nicholas. 1999. *The Cambridge History of Southeast Asia*. Cambridge: Cambridge University Press.

Thomas, Carol. 1993. *Understanding Architecture: Its Elements, History, and Meaning*. Oxford: Westview Press.

Walker, J. H. 2009. "Patrimonialism and Feudalism in the Sejarah Melayu." In *The Politics of the Periphery in Indonesia*, edited by John H. Walker, Glenn Banks, and Minako Sakai, 39–61. Singapore: NUS Press.

Wheatley, Paul. 1983. *Nagara and Commandery: Origins of Southeast Asian Urban Tradition*. Chicago: University of Chicago, Research Paper of Geography.

Yaakob, Usman, Tarmiji Masron, and Fujimaki Masami. 2000. "Ninety Years of Urbanization in Malaysia: A Geographical Investigation of Its Trends and Characteristics." Working Paper, Ritsumeikan: Ritsumeikan University.

10

Urbanism and Planning System in Malaysia

The Historical Roots of Malay Urbanism

In this study, political urbanism is understood as ideas, efforts, forces, processes, forms, and constructions that constitute, concentrate, develop, and undergo urban landscape and settlement. The data of this study rely on the author's fieldwork in the region from 2010 to 2013 and supported with literary studies from various sources concerning the question: What is the relationship between patrimonialism and urbanism in the Malay Peninsula? In dealing with the question, this study investigates the relationship between historical development and setting. Substantially, the study is set forth with a historical review on the transformations and developments of the landscape that matter for and are associated with the urbanism of West Malaysia. The intention of this study is to divulge the ideas, forces, and meanings of the relationship between patrimonialism and urbanism in the Malay Peninsular world.

Until now, there is barely to find surviving written records on the pre-historic Malay Peninsula; the ancient populations were the small numbers of aboriginal people living in the tropical forests as hunter and gatherers (Andaya and Andaya 1982, 9; Bulbeck 2004, 52). The native populations

© The Author(s) 2020
B. Wiryomartono, *Livability and Sustainability of Urbanism*,
https://doi.org/10.1007/978-981-13-8972-6_10

are known as *orang asli* including Semang, Senoi, and Temuans groups. Until today, most of them are hunters and gatherers in various places of the Peninsula forests. They do not develop a community and are with an architecturally organized settlement.

Beyond the native populations' habitat, however, there are the traces and clues of urban settlement. Concerning its architectural relics, the traces lead to Indic civilization that had influenced the formation of states in Kedah areas. The Chinese records from the seventh to ninth century confirm the existence of Hindu-Buddhist states of Langkasuka and Gangga Negara circa the second century (Maguin 2000). Some archeological sites in the Bujang valley show the remnants of Indic town walls made of bricks (Cayron 2006, 32–5). In addition, archaeological evidence between the first century and the seventh century show that the ruling groups in many parts of Southeast Asia had adopted Hindu-Buddhist ideologies, rituals, statecraft, and economics as shown by many artifacts, inscriptions, and images (Bellina and Glover 2004, 68–88).

Historically, the concentration of settlements in the Malay Peninsula can be depicted within three eras: Pre-colonial, colonial, and modern urbanism. Prior to Indic cultural influence, there is hard to find evidence on the existence of state polity and city in the region. The development of indigenous state and city was indispensable from the Indic concept of *negara*; the *Brahmins* and Buddhist monks taught local rulers the state-craft as part of the establishment and consolidation of Hindu-Buddhist influence in the Malay Peninsula and Southeast Asian archipelago (Ooi 2009, xxxv).

Landscape urbanism in Southeast Asia has been unique in terms of culture, society, and architecture that attract scholars from various fields (Berghuis 2012, 204; McGee and Robinson 2011; Rimmer and Dick 2009; Goh and Yeoh 2003; Cairns 2002, 101–20; Evers and Korff 2000); despite their diverse studies, they share something in common regarding the search for something distinctive on urbanism from this region. One important aspect that is considered unique is its long-standing patrimonial political culture. Patrimonial culture is relevant and decisive for planning system and urbanism in this region because it has to do with the way of the political decision-making of human settlement are managed, directed, and operated.

Patrimonialism has been a predominant political culture in Southeast Asia (Case 2014, 88; Budd 2004, 7; Bertrand 2013, 28; Brown 2003, 80, 209; Crough 1978). Patrimonial culture in Malaysia has been shaped and developed with the dyadic system of patron–client relationships. Accordingly, the patrons play an important role in the control of resources while the clients work for the patron as their labors, traders, advisers, and supporters. Urbanism in the patrimonial culture values the necessity for collectivity, conformity, and harmony. In doing so, individual freedom of expression and social justice are perceived as dangers and threats for the integrity and stability of society as well as for the existence and authority of nation-state. In such patron–client relationships, planning system and urbanism do not give a room for creativity from the bottom-up process. Instead, the state and government embody the ruling group as the patron of the ruled populations that culturally has the fatherly power on public policies. In other words, patrimonial urbanism does not take the clients which are the majority of populations, as their citizen but their followers and supporters.

Regarding its political decision-making process, patrimonialism directs and controls local resources based on the masculine values, interests, and preferences of fatherliness. Accordingly, planning and development of urban and regional environment do not give options for other minds and approaches unless for the support and enhancement of the hegemony of the ruling regime concerning unity, integrity, stability, and sustainability of their imagined society and built environment.

The traditional practice of patrimonialism in Southeast Asia is not only the embodiment of economically controlling system of the elites but also elevates and upholds the state as the spiritually performing organization of ceremonial gatherings; the ruler or the King plays the role of conductor and chief of ceremony for the performance (Geertz 1980, 98–120). In Southeast Asia, opposition, contestation, and disagreement against the ruling regime are seen as the disease, disorder, and pollution of its society that socioculturally weakens and disarrays the unity and harmony of the country (Errington 1989, 121).

Demographically speaking, urbanism in this region is characterized by the structure of the center without definite periphery or state without the city in terms of *desakota* (McGee 2002, 31–2). This characteristic might be developed from the pre-colonial times; in ancient times, the land ownership and permanent dwelling were not considered as an important condition for the existence of state, *negara*. Prior to the seventh century, the concentration of settlement is not the case for Southeast Asian states (Wisseman-Christie 1991, 40; Kulke 1991, 12) with his study on Srivijaya suggests; it is possible to infer that there is no 'çity' existed beyond the *kedatuan* Srivijaya in the late of the seventh century; this is probably because most urban life took place on the boats in the river of Musi.

From a historical perspective, unsurprisingly if urban tradition in Southeast Asia did not establish independent 'bourgeois' cities, but it was closely related to the state formation and systems of domination (Evers and Korff 2000, 27). Regarding its peasant demographic structure, the idea of urban settlement in Southeast Asia is not about the concentration of populations in an area but an economically interconnected system of villages. In Javanese tradition, such an interconnection is known as *mancapat* (Bourdier and AlSayyad 1989, 228; Meer 1979, 58–9). In this sense, the establishment of an urban center was not necessarily about the concentration of settlement but the establishment of symbolic power and wealth of the ruling class surrounded and interwoven with the dispersion of commoners' settlement surrounding the seat of the ruler.

Pre-colonial political powers in Southeast Asia configured landscape in Cambodia, Java, and Thailand with an emphasis on the establishment of the spiritual center. As a phenomenon of gathering, ancient urbanism in Southeast Asia was signified by the events of the spiritual congregations or economic exchanges. Architecturally speaking, the structure of settlement does not need a concentric form consisting of core and periphery, but simply a symbolic structure of spiritual unity and a permanent market. Instead of an architecturally concentrated structure, the ancient urban settlement was characterized with a permanent center where the concentration of people for rituals/ceremonies and economic activities took place periodically.

During the course of history of international contacts, pre-colonial Southeast Asian cities show two main characters: the sacred city—*kota*—and the market city—*bandar*—(Brunn et al. 2003, 376); most sacred cities are located in the hinterland areas while the market cities are on the coastal and estuary areas. So far, there are the traces and clues of the sacred city that had been influenced by Indic civilization in Kedah areas. The Chinese records and *batu bersurat*, stone inscriptions, in Perak areas, confirmed the existence of Hindu-Buddhist states of Langkasuka and Gangga Negara circa the second century (Maguin 2000; Jarq-Hergoualc'h 2011). Indeed, in the Bujang Valley, the remnants of Indic town walls made of bricks are found in several archeological sites (Cayron 2006, 32–5). The market city had been established in Malacca in the fifteenth century; the state, *kerajaan*, was originally affiliated with Hinduism then later turn to adopt Islam and became Sultanate of Malacca in 1414 (Ooi 2004, 869).

At the royal capital area, the political dichotomy of domains between the area of the ruling patron and that of the ruled is architecturally apparent that is not only by the layout but also by the quality of building that the center—the areas for the ruler/royal authority, *paduka raja*—is the symbolic orientation of its peripheral settlements where the ruled—*rakyat*—work and live. There is an apt intention for the dualistic categorical differentiation of domain based on the sacred and profane domain. The sacralization of the ruler's residence is architecturally articulated with a landscape design that set the grand mosque of the state as an integrated part of the center. The symbolic dichotomy of the domain between the sacred for the ruling class' residence and the mundane for the ruled area was the underlying principle of spatial order for the Malay urban settlement (Nas and Sluis 2002, 146).

The dualistic spatial ordering system is likely consistent with the binary social order of the Austronesian society (Forth 2009, 191), from which the populations of the Malay and Indonesian archipelago originated. The consequence of this arrangement was the sacralization of the seat of the ruler (Adams and Gillogy 2011, 102). In order to achieve this, religious rituals and ceremonies were liturgically imposed and exercised as part of the state events. The more powerful a center of power is the more mysterious and sacred its form and existence. In order to dramatize this sacredness, the real life of people did not take place in the spiritual center, *istana*, but in the weekly based marketplace, *pekan*. The spiritual center remains useless in daily life.

The conception of the spiritual center had been established during the emergence of Hindu-Buddhist states that was signified by the establishment of their centers called *kota negara*; they were mostly in hinterlands of the Peninsula. The main idea of *kota* is actually designated by the establishment of the residential area with surrounded wall system as the seat of the ruler. The main components of the pre-colonial settlement are the seat of the ruler called *kedatuan* consisting of *istana* (palace) and a square (*dataran*). The center of the state is distinguished from other settlement by the existence of a permanent market, *pekan*, which stands within the walking distance from the palace. Urban life takes place daily in the marketplace. Periodically, state ritual gatherings, ceremonies, and celebrations take place in the square, *dataran*.

Historian Kenneth Hall (2011, 9) notes that Islamic historical records in Southeast Asia are often odds for Western historians; they convey cultural values containing ideas and symbolic meanings which are not easily recognized by Western scholars. The most problematic for this issue lies probably in the fact that local concepts are mostly untranslatable into other languages. The other issues concerning this translation are to realize that amalgamation and evolution of local concepts have been through locally conditioned preferences. One important concept for these preferences that work in the mostly Southeast Asian region is syncretism, *perpaduan*. Accordingly, foreign elements and influences are subject to cultural adaptation, accommodation, assimilation, and harmonization with the existing component toward unity and integrity. The emergence of the concept: *perpaduan* is probably from the practice of patrimonialism. Patrimonial regime prescribes syncretism in terms of harmonious adjustment with local conditions for the sake of the stability and integrity of its hegemony.

Prior to the arrival of the Europeans in the fifteenth century, China and India are important trading partners of states and villages in the archipelago. Since the thirteenth century, the contacts and interactions with Muslim traders from Persia, Arab, and India have developed waterfront vicinities that demonstrated a hybrid of locality and Islamic polity at the state organization and social rank. Islamic influence on the local kingdoms was the establishment of maritime port polities. The Samudera Pasai Islamic Sultanate in Aceh and Malacca state in the Malay Peninsula were among the powerful authorities along the Malacca straits; they established their capital towns based on a hybrid of the Hindu-Islamic polity. Even though Islam introduced people to the egalitarian ritual and practice for solidarity and social justice, the state and its ruler still maintained the religious hegemony that controlled and unified people as a whole.

The acceptance of Islam by local rulers was an interesting point; the ruler takes advantage of Islam regarding its ideological capacity to surpassing multiethnic factions. In addition, Islamic doctrine taught people to be socially egalitarian and obedient with a strong sense of social order and communal responsibility that is useful for the patrimonial state. The outcome of the hybrid of Islamic ideology and Hinduism is *bandar* settlements that still maintains and withhold the sacredness of palace as a symbolic spiritual center; this phenomenon is most notable by the dynamic relationship between the architecture of patrimonial state mosque and palace.

Even though the activities of bazaar were similar to the local traditional market, *pekan*, its multicultural exchanges of the first were more intensive and economically sophisticated than its local counterpart, especially for taxation and currency. Johor Bahru, Muar, and Mersing were the examples of *bandar* settlements in the state of Johor; all these towns are waterfront vicinities. Meanwhile, traditional towns—*kota*—are mostly situated at the river banks and in the hinterland. The toponym of such towns is recognizable with the prefix *kuala* meaning estuaries, such as Kuala Perlis, Kuala Kedah, Kuala Kurau, Kuala Kangsar, Kuala Besut, Kuala Terengganu, Kuala Pahang, Kuala Rompin, and Kuala Selangor. Rivers play an important role for most port towns, *bandar*, and agriculture-based towns, *kota*, in the Malay Peninsula. The waterfront vicinities between the Southeast Asia Peninsula and Sumatra Island have been an international eminence of exchanges between Far-Eastern and Near-Eastern regions since the fifth century (Sheriff 2010, 171–90; Leifer 1978, 6). In terms of *kota*, the use of land for urban area was developed for the economic center of the agricultural community and the symbolic embodiment of spiritual unity. In the Malay world today, an urban settlement is a gazetted area with more than 10,000 populations (Lee 1999, 336). From the observations on the field, the urban settlement is a politically organized vicinity of a populated area that is provided with a permanent market, *pasar* and a municipal status by the state called *majlis perbandaran*.

The important generator of the establishment of *bandar* was the need for trading posts between East and West. Anthony Reid (2000) documents the dramatic transformation of Southeast Asia between 1450 and 1680. It was the period of the establishment and developments of trade posts or *bandar* such as Acehnese Pasai, Malacca, Muar, and Kuala Terengganu. Since most traditional agriculture-based towns—*kota*—were economically self-sufficient, and it did not depend completely on international trades, their role and function remains the same as the symbolic spiritual center, such as Kuala Kangsar of Perak, Pekan, and Kuala Lipis of Pahang. In terms of *bandar*, the use of land is developed for the center of international exchanges between foreign merchants and local-regional traders. Foreign merchants and local-regional traders were the clients of the local rulers.

Patrimonialism and the British Colonial Rule

The involvement of the British Empire in the Malay world dated back to 1786–1826 when they began to establish Penang as one of the Straits Settlements together with Malacca and Singapore. The penetration of the British interest into the Malay Peninsula hinterlands—especially Kedah, Perlis, Kelantan, and Terengganu—was motivated by the mining business of tin and ore, such as in Kinta valleys of Perak and Selangor areas. Since the Pangkor Treaty in 1824, the British colonial rule had established their control over Malay indigenous states of Pahang, Selangor, and Negeri Sembilan, then Kedah, Perak, Kelantan, and Terengganu, multiethnic groups have been populating most urban settlements in the Malay Peninsula either in hinterlands or in coastal areas; the Malay ethnic populations have been mostly predominant in rural areas. Indeed, the proportion of this multicultural society varies depending on its own history. Chinese populations dominate the core of most mining and port towns while the Malay group is dominant in most agricultural settlements. In certain towns such as Penang Georgetown, Taiping, and Ipoh, the Chinese populations outnumber the Malay and Indian group.

Under the British rule, urban settlements were mostly indivisible from mining activities and plantations. Chinese investors, traders, and workers were encouraged to come from the mainland to the Malay Peninsula for the exploration while the British administrators managed the business and taxes (Palmer and Joll 2011, xxi). The unique towns of Malay Peninsula that have been established for ore and tin mining are Ipoh, Bentong, and Kuala Lipis. Even though Ipoh was not established by the British colonial rule, but by the indigenous Malay Landlord: Dato' Panglima Kinta Mohammed Yusuf, the attractive urban architecture of the town was indispensably associated with the flourishing mining business of tin in the region in the early twentieth century. Ipoh is an exemplary synergy of British colonial and indigenous royal town with a well-organized sanitary system.

From the beginning of the twentieth century, Ipoh had been always a multicultural town in the Malay Peninsula. Indian, Chinese, and Australian banks, insurance companies, and trading offices were well represented in this town. Demographically, Chinese Hakka populations have been predominant since the end of the nineteenth century. Besides mining, gambling and opium were the common lifestyles of working class' activities. Needless to say, opium was an important commodity of revenue for the British colonial government (Tak 2009, 201). Nevertheless, the vibrant urbanism of Ipoh was deemed during the Pacific War. Unsurprisingly, in the 1950s Ipoh came back again as the center of entertainment industry of the Malay world. Cinemas, cabarets, cafes, and nightclubs were part of Ipoh's urban culture.

Geographically speaking, the British colonial government maintained and sustained the dividing ethnic groups in their own worlds with the indirect rule (Tay 2010, 2). This includes maintaining the spatial and societal distance between the indigenous ruling class and the native commoners. The presence of Chinese, Arab, Persian, and Indian merchants filled out the gap between the British colonial government and the Malay group. All these ethnic groups: Indian, Chinese, Arab, and Malay populations were kept culturally independent from each other, with their own beliefs, customs, and traditions. The principle of British colonialism is divide and rule (Rappa and Wee 2006, 32). However, the British common laws were in effect for all people concerning crimes and matters of the justice system (Baharuddin 2008, 113).

Since 1957, the administration of land has been the same like in the colonial times, excluding Kuala Lumpur metropolitan area, Labuan, and Putrajaya; these three territories are under the administration of the

Malaysian federal government. The National Land Code (NLC) 1965 is a governing body that establishes a uniform system of regulations concerning land and its tenure in peninsular Malaysia. Accordingly, leases, charges, transfers, and easements are necessarily registered whereas tenancies and liens cannot be registered. Registration of land with the NLC is prescribed for any dealings concerning the uses and transfers of land; no title and no interest will be legitimate without this registration. However, the approval from the local authority for such transfers is required prior to any claim of title and interest of land.

All land in West Malaysia is mostly transferable in terms of a freehold estate or a leasehold property. The tenure of land for a freehold or a leasehold status is commonly from 60 to 99 years. Once the land is registered to the NLC, its record is publicly accessible. In doing so, any interest in the use of land needs to do research on the ownership and history of the land prior to dealings for its acquisition; all information about this is available in the local land office's database. Since 1960 in Johor, the land office, *Pejabat Tanah*, has been an independent state agency and directly responsible to the Sultan. Based on the jurisdiction for land registration, the state of Johor in 2012 has 10 regions, *daerah*, comprising: Johor Bahru, Muar, Batu Pahat, Segamat, Kluang, Kota Tinggi, Pontian Mersing, Kulaijaya and Ledang, three small regions—*daerah kecil*—consisting of: Rengit, Labis, and Pengerang. Under the British colonial rule, *bandar* and *kota* became part of economic hubs of export commodities from plantations: rubber and palm, and mining activities: tin and iron ore. The traditional capitals of indigenous states represented only the symbolic presence of local heritage without any direct access to their resources: land, labor, and capital; the local rulers were nothing but the clients of the British patron without power but only for religious affairs, customs, and traditions.

Patrimonial Culture in Early Post-colonial Urbanism

Until the Malaysian independence in 1957, the Malay Peninsula did not have any comprehensive planning system for physical development; the only planning tools in existence were limited to fire and sanitation regulations in certain urban areas. The physical planning system has been introduced to this country by the British colonial rule in the mid of 1920s. The origin of planning in Malaysia was indivisible from the establishment of municipalities in the British Malay colony on the Malay Peninsula. Prior to those times, there was no urgency to disseminate the ideas and practices of 'Town and Planning' in Britain to the British colonies such as in the Malay Peninsula. Even though the practice of planning for economic development had been working for more than five decades in Malaysia, under the British colonial rule, the first Five-Year Malaysia Plan started in 1950, physical development for urban and regional land has been likely left behind without plans until the 1970s.

The patrimonial patronage of development plans has been exercised within the executive and legislative bodies from federal, regional to a municipal level which is predominantly occupied by the members of UMNO—United Malays National Organization. Since Malaysia's independence in 1957, *Barisan National* or UMNO has been playing its important role in the leadership of planning and development in the country until today. The strong former leader of UMNO such as Mahathir Mohammad is still an influential person in today Malaysian politics; his vision 2020 becomes a hegemonic statement of planning and development in the country (Hilley 2001).

In the Malaysian context, the use of planning system is not only implemented as the tool of development control that has been integrated into the municipal and regional governance since 1976, but it is also the patrimonial legacy of the ruling faction of *Barisan Nasional*. The vision and outlook of the development plans and programs are mostly identical with the UMNO's ideology of development that entails political unity, harmony, and stability. Historically, centralized economic planning system and state intervention have been practiced and shared by most regimes in Southeast Asia along with the beliefs to achieve economic growth, equity, and nationalism (Owen 1999, 142).

The necessity for physical planning was originally not for economic growth, but simply practical for public health. The British rule in the Malay Peninsula had to deal with the similar issues like in Britain, such as sanitation and infrastructure. Land use and infrastructure needed a better management and control in dealing with the growth of populations, especially in their mining towns such as Kuala Lumpur. In 1924, the British rule installed the planning division at its capital city of Kuala Lumpur for coordinating and supervising physical infrastructure, especially for flood control, sanitation, and clean drink water system.

Kuala Lumpur was the first municipality in the Federated Malay States that developed a spatial development plan in 1921. However, the Planning Act for Town and Country was not in existence until 1926. Accordingly, the development and construction of public utilities and infrastructure were subject to municipal approval and supervision. However, the planning section did not take into action for development policy and spatial planning in their respective jurisdiction. The first implementation of the Planning Act for Town and Country was in 1923. Based on the Planning Act, the local government has the authority for setting forth the *General Planning Scheme* or *Structure Plan*. This is an important milestone for the planning practice in Malaysia. In many ways, the act was considered insufficient for providing the municipal and regional authority with the planning tools for the development control of the land.

The acknowledgment of physical and spatial planning as a professional field and practice came into the scope of development control at a municipal and regional level in 1967. It was the time when the Town and Country Planning Act 1967 was issued and disseminated (Sukuran and Ho 2008). The urgency was about coordination for infrastructural development that was necessary at the different level of policymakers, from the Federal, State, to municipal business. After the enactment of the Town and Country Planning Act 1976, every municipality and region in Malaysia is responsible for providing its land with a physical structure plan (Mustafa 2011, 198).

In 1982, the Malaysian government issued the Federal Planning Act that regulates planning matters in special territory under the control of the Malaysian government. The difference between the planning Acts of 1967 and 1982 lies in its implementation; the Federal regulates the areas within the federal territory while the Town and Country Planning Act directs and controls the development works in various states of Malaysia (Town and Planning Act 1976 Malaysia 2005, 97). According to the Street, Drainage, and Building Act 1974, any building development needs to submit a building plan document as a development application to the local authority prior to the construction. The Uniform Building ByLaws 1984 prescribes all technical and administrative requirements for the development application. Based on this, the planning and building section of the local municipality or regional authority will accept or reject the application. Of course, some revisions and amendments should be done for some applications, but some do not have a chance for adjustments, alterations, and modifications. Synchronization with the local interests and needs as represented in its structure plan is the most important part of a successful development application. Negotiations with local authorities are a long way of political process and not likely successful without intervention and help of persons in patrimonial ties.

According to the Environment Quality Act 1974, there are development projects, which are subject to the environmental impact assessment. In this case, such projects need to submit the assessment together with their development application to the local authority. The subdivision developments and industrial facilities belong to these projects. Nevertheless, prior to any development proposal, the acquisition of the land is necessarily at hand. The right of land entitles people to apply a building plan.

The urban characters of the Malay Peninsula are constructed by the concentration of shophouses in a row formation. The Malay, English, Chinese, and Indian characters embellish most urban buildings with their unique ways of expression. Architecturally, modern design based on bricklayer or masonry construction is the features of the structure of its urban scenes. Instead of the uninformative and rigid order of visual landscape character, the Malay towns and cities hold to manage the complexity and variety of form that shape and build an architectural landscape distinctiveness. The infill of the well-ordered architecture of British colonial buildings into the traditional structure of *bandar* enhanced and enriched the scenes of the Malay towns and cities with brick and stone masonry edifices. Even though the use of masonry construction had been developed during the influence of Hindu-Buddhist civilization, its wide-ranging implementation beyond religious and royal buildings happened during the colonial times. Thus, the formative of the Malay urban fabric with permanent structure came into being within this era. The support of the colonial rule to the use of masonry construction was originally for safety and practical purpose in order to minimize the danger of fire in urban areas. Today, the Malaysian building codes have contained such technical requirements and conditions for public safety, health, and comfort in a comprehensive system such as the Uniform Building ByLaws 1984.

In a matter of fact, the due process of development application is more or less a top-down approach. Culturally, local community and developers do not have a platform for dialogues and interactions concerning their interests and needs. Indeed, structure plan is subject to revision and amendment for a certain period of time, every 5 or 10 years. Urban development in the Malay world has been traditionally a top-down policy, meaning the Federation of States has the supreme authority for envisioning, planning, developing, redeveloping, and revitalizing areas according to their opinion and perception.

At the federal level, Malaysian planning work is necessary to go through a political process of the parliamentary system. The chairperson of the National Planning Committee is the Prime Minister of Malaysia. The chief of State Planning Committee is Chief Minister of the state while the head of Local Planning Committee is the local authority; planning committees at each level of authority—from federal, state, to local—are inseparable from the government policy on physical development. In the case of inconsistencies between the plans and the bylaws, the development plans shall prevail. In doing so, the plans are considerably able to overrule the possibly out-of-date regulations. All this is made possible because of the fact that some regulations have not been reviewed and adjusted to deal with current issues and circumstances. In order to be efficient, the state planning committee is established at the federal and municipal level (Mustafa 2011, 199–200). The room for the improvement and rehabilitation of policies and plans is on the planning committee. However, for its legislation, the plans and policies still need a solid approval from the state council. Even though the authority of planning committee is politically strong, its plans and policies are not open for appeal and plead; the Planning Act 1976 allows the local authority to set up an Appeal Board, *Lembaga Rayuan*. Accordingly, development applications from private sectors have to comply strictly their plans: for land use, density, building, and infrastructure with the given structure plan and local plan.

In the periods between the 1980s and 1990s, physical developments in Peninsular Malaysia experienced massive activities that transformed the landscape from the natural to the built environment. Political pressures for the physical development control imposed a dual need: For greater technical proficiency in planning and for politically enhanced control over planning (Rudner 1994, 201). The hierarchy of decision-making in the planning process is strongly regulated by the federal, state, to the local authority, from the National Physical Plan, State Structure Plan, to Local Plan. The physical development at any level of planning is not only to comply with the Planning Act but also to be in concert with the respective National-Five Year Plan; the First Malaysia Plan was for the period between 1966 and 1970.

According to Act 1976, all municipalities and regions have to establish councils, *majelis perbandaran*, and *majelis daerah*. City and county council members are ordinary citizens and residents of each area within the municipality. They were selected and represent the local interests and aspirations. Indeed, public participation in the planning process in this country is not comparable with the culture of community involvement in democratic countries, given that the planning authority of Malaysia provides for public participation in two stages: to prepare and complete the structure plan stage.

According to the Planning Act, community participation is an integrated part of the process. However, critical position and disposition of the plan are unfamiliar and unlikely encouraging in the Malaysian society. Instead of the opposition and objections, the pressure for integration, cohesion, and uniformity among federal, state, and local policies came to be the first priority. Even though the public exposition for a new structure plan is the part of the planning process, public participation in this country is socioculturally not comparable with the democratic involvement of local communities in other democratic countries. Malaysian planning authority gives such public participation in two stages: for the preparatory and the finalize stage of the structure plan.

Seemingly, Malaysian multiethnic communities are not yet ready for being actively engaged in the political process for the planning system. Culturally, in Malaysia, the patrimonialism of government has been an integrated part of the governing style regardless of who is in power (Pye and Pye 2009, 261). Instead, of oppositions and objections to a plan, the pressure for integration, cohesion, and uniformity between federal, state, and local policies comes to be the first priority. Criticism and contra-arguments will be managed with a persuasive deliberation toward a conformist resolution.

Unlike in the pre-colonial period of trades between 1530 and 1630, contemporary urban settlements in the Malay world, either *kota* or *bandar* comprises several units of urban communities with access to international trade activities (Grime 1996, 719–20). The different character between *kota* and *bandar* comes to the end. While pre-colonial urbanization along the waterfront areas of Southeast Asia was populated by 5% of the whole populations living in many kota and bandar (Yap 2012, 20), 73% of Malaysian populations live and work in urban areas in 2012 and 2013 (The World Bank Report 2014). This most recent demographic figure shows the dramatically growing Malaysian urban society. In order to achieve its economic sustainability, Malaysia has begun to implement programmatic knowledge-based urban development. Since Prime Minister Najib Razak took office in 2009, the dependency on traditional resources—land, labor, and capital—was gradually transformed to by a programmatic scheme for the reliance on economic knowledge, with the improvement of multimedia cyberinfrastructure and the wide range utilization of green renewable energy. One important thing of the goals concerning urban development based on the knowledge-based economy is the formation and establishment of an attractive cosmopolitan urban society with highly skilled populations in various fields.

Patrimonialism and New Town Development: Putrajaya and Iskandar

The implementation of modern planning for urban development did not come into the horizon of the Malay Peninsula until the 1990s. A new town development in a full system is the model for this implementation, such as Putrajaya and Syah Alam (King 2008, 111–2). The plan and development of Putrajaya were developed based on the vision of Datuk Mahathir Muhammad, Prime Minister of Malaysia from 1981 to 2003. The vision of Putrajaya is the capital city of the Malaysian nation-state with modern garden and intelligent high-tech infrastructure. Since the nation-building in Malaysia is not an easy thing to do, the establishment of Putrajaya is an architecturally imagined construction of Malaysian citizenship in the global community.

In contrast to the pre-colonial and colonial urban settlements that grew and developed in response to the market forces of economic and socio-cultural activities, Putrajaya is an artificially established city by a government political will; it is an architecturally planned urbanism with its core activities for government business with its residential support (King 2008, 128–68). The vision of the city is an actualization of the Malaysian Outlook, Wawasan Malaysia, 2020 (Doolittle 2005, 105–6). Green landscaping and modern infrastructure characterize the feature of its obvious urban appearance. Even though the significant number of Malay, Chinese, and Indian populations constitute the country's multicultural demography, their ethnic heritage is architecturally not well represented as the Malaysian urban reality.

Modern building forms mostly dominate the scene of urban architecture in Putrajaya; the experimentation of hybrid between Islamic and modern element seemingly becomes the chief idea, theme, and direction of this imagined city plan and design. What lies behind this entire grandiose scheme is likely nothing but the fact that Putrajaya lacks an architectural idea for a nationally unifying system of diverse cultures and traditions. Despite its technological sophistication of building utilities and structure, Middle Eastern architecture as a model shows only the helplessness of the search for architectural identity, authenticity, and integrity based on its own resources. Do this monumental layout and its landmarks represent the Asian societies' values, which are more harmonious, more morally upright, more efficient, and disciplined than those of the West (Ang 2001, 6).

Why should the architectural forms in Putrajaya and Iskandar City emulate the Middle Eastern architecture; domes, arches, abstract geometric patterns play an important role in its most architectural forms. Self-doubt and inferiority of the worldwide power of the West and the Chinese economic supremacy could be the underlying motif of Malaysian architectural identity of Putrajaya's and Iskandar city's important public buildings with the Middle Eastern world; the monumental scale of avenues and buildings shows the rupture of traditional Malaysian life and lacks architectural coherence with alienated pedestrian convenience. This case reminds us of the criticism of Niemeyer's Brasilia (Holston 1989).

Nevertheless, all these questions regarding cultural identity probably find the answer in the imagined Malaysian nation-state, in terms of national political agenda of the United Malays National Organization, UMNO, under the direction of Mahathir Mohammad as former Prime Minister between 1981 and 2003.

Morphologically speaking, the city is developed with a strong axial system of monumental landscape juxtaposition. The whole outline is developed with a strong axial framework running from The North, where the Prime Minister office and the complex of executive offices are, to the South, where the judicial institutions and the International Convention Centre are located. As a whole composition, the architectural configuration of Putrajaya's layout demonstrates a monumental design with grandiose edifices.

The spatial orientation of the capital complex is to signify the direction to Qiblah or Mecca. From this direction, the central boulevards, *persiaran perdana*, run to distribute the traffics to various institutions of the Malay Federated states. Islamic spirit underlies the layout and its landmarks with Middle Eastern architectural identity in a tropical setting. Environmentally speaking, the sustainability of Putrajaya City is arguably for the energy conservation and for the environmental protection of the existing habitat at the lake areas and its Wetlands (Knaur and Hitam 2010, 168–9).

Regarding its patrimonial political culture, the whole process of planning for Iskandar Malaysia Project is a top-down approach. The initial idea of the planning was from the Malaysian government under the former Prime Minister Abdullah Badawi in 2006. As its predecessor at the Putrajaya development, Iskandar Malaysia is a modern project without inviting and attracting the participation of local communities in Johor Bahru areas for its planning process.

Iskandar Malaysia or Iskandar Development Region is a new development project in South Johor. Iskandar Regional Development Authority administers the project comprising 2217 square km of land. The project is a regional growth conurbation, which started in 2006; the state of Johor is the main sponsor and key player of development. The purpose of the development is to establish a Multimedia Super Corridor of economic growth. The approach to the project is a rational approach with a long-range comprehensive plan. With such a project, Iskandar Malaysia can be seen as an insistent project to challenge Singapore's reputation as a well-established financial and economic hub in Southeast Asia (Tey 2011, 131). The land is relatively an empty area with less-utilized terrain for settlements and agriculture. In short, the site is perfect for new development with fewer obstacles from the possible displacement of people living in the area. Administratively, the whole area of development has been integrated into a territory called Nusajaya. In doing so, a newly established authority will administer regional and handle all planning and design matters.

The necessity for developing the land for Iskandar City is obviously in response to the economic growth of the region generated by its Southeast Asian growth center of Singapore. Indeed, Iskandar City is by no means the backyard of Singapore's economic growth. Programmatically, Iskandar City is not only the administrative headquarters of Johor state, but also a regional growth center of West Malaysia. Besides its residential, commercial, and industrial development, the focus of the land improvement in Iskandar Malaysia project is to establish an educational city with its strength in the business of health care and tourism.

Iskandar Malaysia or Iskandar Development Region is a new development project in South Johor. Iskandar Regional Development Authority administers the project comprising 2217 square km of land. The project is a regional growth conurbation, which started in 2006; the state of Johor is the main sponsor and key player. The purpose of the development is to manage the future needs and resources for economic growth in the region. The approach to the project is a rational approach with a long-range comprehensive plan. Even though the future growth of the Southeast Asian region is uncertain due to the fluctuation of the global economy, there is no a significant and well-thought resistance or opposition to Iskandar Malaysia mega project in the context of open professional and political discourse.

Regarding its scale and volume of the project funding, such a development plan should be a necessary part of the democratic process for a nation-building where the socioculturally segregated ethnic groups have the chance and opportunity to work together as a nation. The Malaysian way of planning and development is more or less paternalistic in its practice, with the concentration of power at the leaders of the ruling establishment. Political father figure plays an important role in the real politics in inspiring, motivating, and mobilizing people to do something great for the country. Contestation and opposition will be seen as shameful and disrespect against the founding fathers of the country.

The site of Iskandar Malaysia project is relatively vacant and with less-utilized terrain for industries and agriculture. However, the site is perfect for new development with fewer obstacles from the possible displacement of people living in the area. Administratively, the whole area of development has been integrated into a territory called Nusajaya. In doing so, a newly established authority will administer regional and handle all planning and design matters.

The necessity for developing the land for Iskandar City is obviously in response to the economic growth of the region generated by its geographic relation to Singapore. The vision of Iskandar city is the twenty-first-century metropolis of Malaysia with highly developed technology for its structure and infrastructure that provides high-quality standards for the live-work-play of a knowledge-based society. In dealing with this vision, the industrial productivities and services of Iskandar city are supported by international airport of Senai and seaport of Tanjung Pelepas with 1,800,000 square m container area, free-zone, marina facilities, and procurement.

As a whole productive center, Iskandar city is provided with world-class higher educational institutions, health cares, amusement parks, trade centers, various residential units, and green industrial estates. Nevertheless, Iskandar City development is to strive its best that is to challenge Singapore's industries and services. On the contrary, it is an economic zone development in the Southeast Asian region, probably similar to the Shenzhen industrial zone. Programmatically, Iskandar City is not only the economically productive zone of Johor state but also a regional growth center of West Malaysia. In a matter of a fact, Iskandar city is one of the examples of urban development by state sponsorship with massive investment and organization on earth today. In democratic and well-developed industrial countries in North America and Europe, such a development is almost impossible because it will be not executable within the terms of political administration for maximum 8 or 10 years.

Programmatically speaking, the focus of the land development in Iskandar Malaysia project is to establish a sustainable growth region of Southeast Asia with populations for initially 1.4 million and 3.0 million in 2015 (Joeman 2013). Besides its various zones of development for residential, commercial, and industrial activities, the project strives to provide innovative facilities for businesses, educations, health care, entertainments, and recreations. Indeed, the problem of this development is obviously about workforces. How could such an ambitious development project be realistic in dealing with human resources that are able to run the project from its early stages to its operation and maintenance? Johor or Malaysia is not Shenzhen or China in terms of populations. Until now, the Iskandar Malaysia project did not show a plan on how to develop human resources in a viable demographical condition; who will work and live in the region in the coming decades? The uncertainty of the world economy overshadows the future of the project with many questions.

Concluding Remarks

The relationship between patrimonialism and urban landscape development in Malaysia has been going through from three periods of political regime, under the pre-colonial, colonial, and post-colonial regime. The process of landscape configuration and transformation, as well as development, has been politically far from a democratic process. Patron–client ties with the ruling regime direct and manages the urban landscape development from its very beginning to the practice and sustenance of the patrimonial control over economic resources: labor, land, and capital. In this patrimonial framework of decision-making, the planners and designers of the built environment have no much option but as the tools of patrimonialism to actualize the regime's imagined society and built environment. Unless its very idea and program are in alignment with the political agenda of the ruling regime, the ingenuity of artistic and intellectual expression is not likely to welcome. Planning and design outcome under patrimonial regimes do not open options for citizen participation that build, develop, and sustain the sense of community with strong identity between people and place.

References

Adams, Kathleen, and Kathleen Gillogy. 2011. *Everyday Life in Southeast Asia*. Bloomington, IN: Indiana University Press.

Andaya, Barbara Watson, and Leonard Y. Andaya. 1982. *A History of Malaya*. Houndmills: Macmillan.

Ang, Ien. 2001. *On Not Speaking Chinese: Living Between Asia and the West*. London and New York: Routledge.

Baharuddin, Syamsul Amri. 2008. "Competing Domains of Controls: Islam and Human Rights in Malaysia." In *Islam and Human Rights in Practice, Perspectives Across Ummah*, edited by Shahram Akbarzadeh and Benjamin McQueen, 1081–117. London and New York: Routledge.

Bellina, Berenice, and Ian Glover. 2004. "The Archaeology of Early Contact with India and the Mediterranean World from the Fourth Century BC to the Fourth Century AD." In *Southeast Asia from Prehistory to History*, edited by Ian Glover and Peter Bellwood, 68–88. Oxfordshire: RoutledgeCurzon.

Berghuis, Thomas. 2012. "Architecture." In *Cultural Sociology of the Middle East, Asia and Africa: An Encyclopedia*, edited by Andrea Stanton and Edward Ramsamy, 204–9. London: Sage.

Bertrand, Jacque. 2013. *Political Change in Southeast Asia*. Cambridge: Cambridge University Press.

Bourdier, Jean Paul, and Nezar AlSayyad. 1989. *Dwellings, Settlements and Traditions: Cross-Cultural Perspectives*. Lanham, MD: University Press of America.

Brown, David. 2003. *The State and Ethnic Politics in Southeast Asia*. London: Routledge.

Brunn, Stanley D., Jack Williams, and Donald J. Zeigler. 2003. *Cities of the World: World Regional Urban Development*. Lanham: Rowman & Littlefield.

Budd, Eric N. 2004. *Democratization, Development and the Patrimonial State in the Age of Globalization*. Lanham, MD: Lexington Books.

Bulbeck, David. 2004. "Dong-Son Culture in Southeast Asia." In *A Historical Encyclopedia from Angkor Wat to East Timor*, edited by Kiat Gin Ooi, 428–39. Oxford: ABC-CLIO.

Cairns, Stephen. 2002. "Troubling Real Estate: Reflection of Urban Form in Southeast Asia." In *Critical Reflections on Cities in Southeast Asia*, edited by Timothy Bunnell, Lisa Barbara Welch-Drummond, and Kong Chong Ho, 101–20. Singapore: Brill & Times Academic Publishing.

Case, William. 2014. *Routledge Handbook of Southeast Asian Democratization.* London: Routledge.

Cayron, Jun G. 2006. *Stringing the Past: An Archaeological Understanding of Early Southeast Asian Glass Bead Trade.* Diliman Quezon City: University of Philippines Press.

Crough, Harold. 1978. *The Army and Politics in Indonesia.* Ithaca: Cornell University Press.

Doolittle, Amity Appel. 2005. *Property and Politics in Sabah, Malaysia: Native Struggle over Land Rights.* Seattle, WA: University of Washington Press.

Errington, Shelley. 1989. *Meaning and Power in Southeast Asia.* Princeton, NJ: Princeton University Press.

Evers, Hans-Dieter, and Rudiger Korff. 2000. *Southeast Asian Urbanism: The Meaning and Power of Social Space.* Muenster: LIT.

Forth, Gregory. 2009. *A Tale of Two Villages: Hierarchy and Precedence in Keo Dual Organization, Flores Indonesia.* Canberra: ANU Press.

Geertz, Clifford. 1980. *Negara: The Theatre State in the Nineteenth Century Bali.* Princeton: Princeton University Press.

Goh, Robbie G. H., and Brenda S. A. Yeoh. 2003. *Theorizing the Southeast Asian City as Text: Urban Landscapes, Cultural Documents, and Interpretive Experiences.* Singapore: World Scientific Publishing.

Grime, Charles E. 1996. "Indonesian, an Official Language of Multi Ethnical Nation." In *Atlas of Languages of Intercultural Communication in Pacific, Asia, America*, edited by Stephen Adolphe Wurm, Peter Mulhauser, and Darell T. Tyron. Berlin: Walter de Gruyter.

Hall, Kenneth R. 2011. *A History of Early Southeast Asia: Maritime Trade and Societal Development 100–1500.* Lanham, MD: Rowman & Littlefield.

Hilley, John. 2001. *Malaysia: Mahathirism, Hegemony and the New Opposition.* London and New York: Zed Books.

Holston, James. 1989. *The Modernist City: An Anthropological Criticism of Brasilia.* Chicago: University of Chicago Press.

Jarq-Hergoualc'h, Michel. 2011. *The Malay Peninsula: Crossroad of the Maritime Silk Road 100 BC–1300 AD.* Leiden: Brill.

Joeman, Boyd Dyonisius. 2013. *Iskandar Malaysia Smart City Framework.* March 21–22. http://hls-esc.org/documents/4hlsesc/2C%20-%20Iskandar.pdf.

King, Ross. 2008. *Kuala Lumpur and Putrajaya: Negotiating Urban Space in Malaysia.* Singapore: NUS Press.

Knaur, Hadev, and Mizan Hitam. 2010. "Sustainable Living: An Overview from Malaysian Perspective." In *Towards a Liveable and Sustainable Urban Environment, Eco Cities in East Asia,* edited by Lye Liang Fook and Chen Gang, 159–78. Singapore: World Scientific.

Kulke, Hermann. 1991. "Epigraphical Reference of City and State in Early Indonesia." *Indonesia* 52 (SEAP): 3–22.

Lee, Boon Thong. 1999. "Emerging Urban Trend and Globalizing Economy in Malaysia." In *Emerging Cities in Pacific Area,* edited by Fu-Chen Lo and Yeong, 336–48. Tokyo: UNU Press.

Leifer, Michael. 1978. *Malacca.* Dordrecht: Brill.

Maguin, Pierre-Yves. 2000. "City-States and City-State Cultures in pre 15th Century Southeast Asia." In *A Comparative Study of City-State Cultures: An Investigation,* edited by Mogens Hermann Hansen, 409–30. Copenhagen: University of Copenhagen.

McGee, T. G. 2002. "Reconstructing 'The Southeast Asian City in an Era of Volatile Globalization'." In *Critical Reflections on Cities in Southeast Asia,* edited by Tim Bunnell, Lisa B. W. Drummond, and K. C. Ho, 31–53. Singapore: Brill & Times.

McGee, T. G., and Ira M. Robinson. 2011. *Mega Urban Regions of Southeast Asia.* Vancouver: UBC Press.

Meer, N. C. Van Setten van der. 1979. Sawah Cultivation in Ancient Java, Aspects of Development during the Indo-Javanese Period 5th century. Oriental Monograph Series No. 22. Canberra: ANU Faculty of Asian Studies.

Mustafa, Maizatun. 2011. *Environmental Law in Malaysia.* AH Alphen aan den Rijn: Wolters Kluwer.

Nas, Peter J. M., and Reynt J. Sluis. 2002. "In Search of Meaning." In *The Indonesian Town Revisited,* edited by Peter J. M. Nas, 130–46. Muenster: LIT.

Ooi, Keat Gin. 2004. *Southeast Asia: A Historical Encyclopedia, from Angkor Wat to East Timor.* St. Barbara, CA: ABC-CLIO.

———. 2009. *Historical Dictionary of Malaysia.* Lanham: Scarecrow.

Owen, Norman G. 1999. "Economic Social Change." In *Cambridge History of Southeast Asia from the World War II to the Present,* edited by Nicholas Tarling, 139–98. Cambridge, UK: Cambridge University Press.

Palmer, David, and Michael Joll. 2011. *Tin Mining in Malaysia 1800–2000: The Osborne & Chappel Story.* Kuala Lumpur: Perpustakaan Negara.

Pye, Lucian W., and Mary W. Pye. 2009. *Asian Power and Politics: The Cultural Dimensions of Authority.* Cambridge, MA: Harvard University Press.

Rappa, Antonio, and Lionel Wee. 2006. *Language Policy and Modernity in Southeast Asia: Malaysia, the Philippines.* Singapore: Springer.

Reid, Anthony. 2000. *Southeast Asia in the Early Modern Era: Trade, Power, and Belief.* Ithaca, NY: Cornell University Press.

Rimmer, Peter James, and Howard W. Dick. 2009. *The City in Southeast Asia: Patterns, Processes and Policy.* Singapore: NUS Press.

Rudner, Martin. 1994. *Malaysian Development: A Retrospect.* Montreal: McGill-Queen University Press.

Sheriff, Abdul. 2010. *Dhow Cultures and Indian Ocean: Cosmopolitan, Commerce, and Islam.* New York: Columbia University Press.

Sukuran, Mohammad, and Chin Siong Ho. 2008. "Planning System in Malaysia." Skudai Johor Bahru: Toyohashi University of Technology-UTM Seminar of Sustainable Development and Governance.

Tak, Ming Ho. 2009. *Ipoh.* Ipoh: Perak Academy.

Tay, Eddie. 2010. *Colony, Nation, and Globalization: Not at Home in Singaporean and Malaysian Literature.* Hongkong: Hongkong University Press.

Tey, Tsun Hang. 2011. "Iskandar Malaysia and Malaysia's Dualistic Political Economy." In *Special Zones in Asia Market Economies*, edited by Connie Carter and Andrew Harding. London: Routledge.

Town and Planning Act 1976 Malaysia. 2005. "Planning Act 1976." In *Compendium of Land Use Laws Development*, edited by John Nolon, 97–111. Cambridge, UK: Cambridge University Press.

Wisseman-Christie, Jan. 1991. "States Without Cities: Demographic Trends in Early Java, Indonesia." *Indonesia* 52: 23–40.

Yap, Kioe Sheng. 2012. "The Challenges of Promoting Productive, Inclusive and Sustainable Urbanization in Southeast Asia." In *Urbanization in Southeast Asia: Issues & Impacts*, edited by Kioe Sheng Yap and Moe Thuzar. Singapore: Institute of Southeast Asian Studies.

11

Urban Intentionality and Global Urbanism: Toronto as a Case Study

Urbanism and Consciousness

Intentionality is the mental states and affairs by which people are directed spontaneously to engage and experience in the state of affair of the world. Because of intentionality, people are able to deal with live, work, and play in certain ways and manners that establish, develop, and maintain the urban life-world. Until recently, urban studies remain oblivious to the phenomenon of conscious awareness in terms of intentionality. Intentionality concerns the ontological question of urban conscious awareness in the context of urban experience; it is about how to experience the urban life-world that exists and presents.

Urbanism is the thing of thinking, planning, designing, managing, building, developing, and maintaining an urban phenomenon; it is an intentional entity. Urbanism contains the intentionality of urban existence and presence. In daily life, the presence and existence of a thing are only accessible and experienceable through intuition. Since intuition works and operates at the conscious awareness, this cognition is the condition of mental states or events when humankind understands and able spontaneously to do something in the context of engagement or involvement

© The Author(s) 2020
B. Wiryomartono, *Livability and Sustainability of Urbanism*,
https://doi.org/10.1007/978-981-13-8972-6_11

in the world. Practically speaking, practical conscious awareness enables people intuitively to do their routines and chores without any thought.

What is the relationship between urbanism and the human mind in relation to intentionality? The question leads us to go into the rudimentary condition of human existence as social being in the urban context; what and why they are and do work, live and play together as members of a sociopolitically organized community with a geographically concentrated settlement for their sustainable well-being and growth. Theoretically speaking, intentionality is necessarily considered as an essential part of the relationship between mind and the world because it is the way of being in the whole system of beings (Heidegger, *Being and Time* 1927/2010, Chapter 9; Tonino 2003).

Despite the fact that urbanism is ideologically manifold, it is an indispensable part of collective conscious awareness for practicality that works spontaneously and instantaneously as mentally driving force for experiencing, doing, and developing their urban world as their home. This practical conscious awareness is considered as urban intentionality. Historically speaking, the concept of intentionality leads us to Edmund Husserl's phenomenology since his *Logical Investigations I & II* and *Ideas* (Husserl, *Logical Investigations*, 1901–1902/2001; Husserl, *Ideas, General Introduction to Pure Phenomenology*, 1962/2002/2014). Accordingly, intentionality is the conceptual and referential thing of mental states or events in engaging and experiencing the world (Searle 1983, 231–61); the world is a whole system of beings in manifold form of institutions that happens historically; house, neighborhood, county, town, city, country, and global community are the worlds.

The urban life-world in this sense is considered as the thing toward which the collective mental states or events of people are directed and care for its presence and existence. Presence in this sense is understood as what it is experienceable for sensory data. While existence is the capacity of humankind—as a sentient being—to stand out from the trap of routine temporality into future. Urban practical conscious awareness is the mental ability of humankind to engage and experience the world as a collectively shared system of the dwelling. Based on this socially shared cognition, people are directed and moved spontaneously into the manifold activities of urban production, service, and leisure that bring about the economically unified system of livability and sustainability in highly concentrated buildings and populations; we call these mental states and events as urban intentionality. In other words, urban intentionality is something that has the transcendental target of mental acts toward and about which urban dwellers subconsciously share something in common about their existence and presence. Accordingly, urban existence is about the sustainable reality of collective work, live, and play together as a sociopolitically organized community in a highly concentrated settlement that is and grows economically, culturally, and spiritually. Urban presence concerns about the vibrant livability of people in urban public places that demonstrates safe, healthy and enjoyable interactions.

Urban conscious awareness contains vibrantly crowded intentionality that directs, drives, and integrates people's movements and activities into urban socioeconomic and communication network. This intentionality is believed to have specific human brain cycles, vibrations of the landscape, and their built environment that plays and works as unique ambiances of locality. Such ambiances attract, direct, and stimulate their brain activities at certain frequencies; the unity of these manifold vibrations constructs and develops unique proximity, diversity, connectivity, and livability that matter for the presence of urbanity. To a certain extent, these vibrations make them feeling relief from their sense of self, and free from logical constrictions of daily routines, and enjoyable flow into the streams of the crowd in public places. Urban intentionality—like a general intentionality—has a holistic network of collective conscious awareness so that urban dwellers are able to mingle, engage, and participate spontaneously into the flow of crowd or daily worldly affairs in public places.

In daily life, urbanism is experienced by the human mind at the mental states and events of conscious awareness. Casual interactions in the anonymous crowd with conscious awareness have been described by German sociologist Georg Simmel—from his 1903 essay: The Metropolis and the Mental Life—as the emotionally fascinated experience of urban life but with blasé attitude (Simmel 1950). Simmel speaks of the mental life of blasé in the context of individual preservation against enormous social forces of urban public places. Accordingly, despite the intensification of nervous stimulation of metropolis, urban people enjoy individual freedom in their anonymity. This freedom includes the aesthetic experience of the public sphere. Walter Benjamin in *Arcade Projects* invents the complexity of experience in modern urbanism that involves reciprocity between urban life and arts, especially literary and visual art (Benjamin, *The Arcades Project,* 1999). Benjamin presents his analysis of Paris urban experience based on the allegorical intuition of *Le Cygne* and *Le Fleurs du mal* by Charles Baudelaire (1821–1867). Benjamin's personal account on urban experience is described in his 1932 biographical sketch: *A Berlin Chronicle,* in this writing, Benjamin presents a recollection of photographic memory into a synergic literary work of descriptive narrative, memoir, montage, essays, and aphorism. All this in the search for urban reality based on personal experience.

Simmel and Benjamin share something in common concerning the urban reality of the modern capitalist world in the early twentieth century. Accordingly, urbanity in modern cities is inescapable from the dynamics of socioeconomically intertwined reciprocity between capitalist-technological production and social life of people. The ancient question remains the same: How urbanism is able to uphold human decency? This question is valid for any city on the globe. Toronto is one of the North American most populous urban centers that have to deal with the fundamental question of its existence and presence.

Based on 2016 Statistics Canada, the City of Toronto, Ontario is the largest urban center in Canada and home of 2.731 million populations; statistically speaking, half of the total populations were born outside the country (Statistics Canada 2009-11-20). The concentration of populations and buildings occupies about 630 square kilometers; between 2001 and 2006, four in every ten newcomers to Canada settled in Toronto. What makes Toronto attractive to people around the globe? There are numerous reasons for this from personal to general preferences. However, from the perspective of the spiritual perspective of urbanism, Toronto stands out to attract people's soul from everywhere on the globe because of its unconditional vibrations? How can Toronto qualify for possessing such vibrations? The proof of this is necessarily well articulated in the mission statement of town or city as set forth in the official plan or master plan or in other long-range planning documents. A well-formulated mission statement of a town or a city represents and reflects a profound understanding of its urbanism in respect of its identity with its potentialities and resources.

The conscious awareness of an engagement in the worldly states of affairs is spontaneous but intentional in terms of aboutness and directedness so that people are socially able to work together purposely for the manifold activity of productions, services, and leisure. This spontaneous conscious awareness for purposely activities we call it intentionality; this is the faculty of mind for understanding that enables humankind to navigate, orient, engage, participate, occupy, listen, read, drive, and move with the mentally relaxed condition. In the mental states and events of understanding, people are non-verbally communicating and interacting through body language, gestures, and signs so that they are able to work together in public places. These mental states or events are within the brain cycle of 8–13 Hertz. This cycle is known as an alpha wave. There are other cycles called delta, theta, beta, and gamma; the names of the cycle coined by German neurologist Hans Berger in his 1929 (Berger 1929; Buzsaki 2006, 4). Accordingly, the mental states and events of understanding enable people to deal with their urban daily life on streets, workplaces, and leisure areas.

Casual interactions do not need spoken language but people understand each other that enable them to establish the orderly structured system of habitation and routine. In doing so, the sense of order has been prescribed non-verbally prior to any normative form of regulations and controls. The utmost purpose of these normative contents is to fill the spiritual need of people; the content of urban intentionality is urbanity. Without this content, urban existence and development do not have any sense. Urbanity is about human decency that is established, developed, and upheld as the synergy of manifold ideas, efforts, forces, activities, dreams, hopes, feelings, and desires of people who believe and participate in the making of urban settlement. In short, urbanity is about civility.

The phenomena of urban habitation show how routine activities are potential to construct the sense of home in a place or an environment. All this has something to do with the feeling of safe and free from their fear of the unknown. What do people feel their city is their home? Home is not simply the place to stay, work, and play in peace and dignity, but also the environment from where they can stand out and grow. Urbanism is a phenomenon of human cultivation through concentration of settlement. Under the notion of cultivation, all resources—material, labor, energy, fund, and time—are organized, engineered, developed, handled, mobilized, and distributed into a system of production and service that is to cultivate what is to be human.

In the course of history, the purpose of urbanism is inescapable in the search for being human in manifold forms, styles, manners, and ways. What makes urban citizens different from others is their daily experience of their populations and built environment. Historically speaking, the sense of home in an urban context is distinguished from other places because of their experience of proximity, connectivity, diversity, and livability of people and places. Urban planning and policy are outlined, designed, drafted, and enacted to deal with the issues about this experience at the normative level.

The problem of planning and policy lies in its nature in directing and framing principles and areas of concern that work categorically for the public good. Such a normative system of principles and areas of concern should be conceptually wide-ranging and everlasting. Consequently, such a system is necessarily drawn from the collectively shared system of ideas, desires, hopes, and expectations of urbanism; this system is cognitively framed with the notion of urban intentionality. Despite the manifold of geographical, historical, and ideological contexts of urbanism, urban intentionality is believed to work as categorical principles and areas of concern for urbanity.

Under the concept of urbanity, all efforts, forces, and activities in urban context are organized, managed, directed, developed, and sustained toward the unity of vibrant well-being and decency; this is urban intentionality that brings about the necessity for proximity, diversity, connectivity, and livability of urban phenomenon to come into purposive integration. Without urbanity as intentionality, urban planning and policy do not have an unconditional purpose that is everlasting and resourceful.

The integration of vibrant well-being and decency is considered as the collectively spiritual necessity for urbanity. Urban intentionality manages, directs, and navigates the collective conscious awareness of urban dweller toward the sense of urbanism that is to establish, develop, uphold, and sustain the liveable well-being and decency of humankind with a sustainable environment. In the daily world of urbanism, all this intentionality is necessarily incorporated into the vision of urbanism for town or city. The vision is considered as the categorical source of goals for planning, design, policy, and implementation. Since urban intentionality is referential in the context of its people and their local resources, identifying the authentic distinctiveness of an urban center is indispensable. This identity is not simply about the geographic, demographic, and historic profile and potentiality. Rather, this identity is necessary to dismantle the spiritual character of the urban phenomenon; this includes the geomagnetic forces, astrological constellation, underwater streams, chemical processes in the environment, memories and stored vibrations; all these are experienced by people who live, work, and play in a certain location with certain emotional responses. Most of the modern people ignore and do not pay attention to such responses. However, whatever the forces are, what does really matter for an urban center is the question: What kind of a world this city is? What does this city make different from others, for example, in relation to specific issues such as gay marriage, climate change, and Islamic phobia?

The urban intentionality of the City of Toronto has been generally comprehended within its official plan's vision on healthy future with the conducive and attractive environment for investment and high quality of life (The City of Toronto 2002, 17). Despite a comprehensive identity of a city is not an easy task and probably almost impossible because of its manifold potentialities and meanings, the search for this distinctiveness is necessary that is in order to acknowledge, recognize, and appreciate its presence. Incorporating the urban intentionality in a normative document is regarded as an attempt to incorporate the collective conscious awareness of a world with specific physical and non-physical properties and potentialities.

There are other attempts to grasp the urban intentionality of Toronto from various perspectives, such as from its liberal tradition of its founders (Lemon 2008, 242–94) and newspapers (Mackintosh 2017), from its multicultural school world (Vipond 2017), and from pragmatic governmentality (Redway 2014), from its agility of its historical development (Amstrong 1988), from the historical daily world (Levine 2014), from its architectural ingenuity (Stanwick et al. 2007), from its innovative planning tradition (Sewell 1993), and from artistic and cultural inclusivity (Goldberg-Miller 2017). From these various points of view, urban intentionality is about signature, personality, characteristics, property, and profile of being. The urban phenomenon is not a permanent object but a living thing that exists and grows.

The relationship between urban intentionality and its phenomenon lies in the care of its inhabitants; Heidegger describes care as the structure of being (Heidegger, *Being and Time* 1927/2010, 177–78). Urban planning and policy are the manifestations of this care. In the practical world of urbanism, the care is not only about environmental conservation and preservation, as well as maintenance of public utilities and services, keeping historical sites, upholding urban rituals and traditional events. The care for urbanism includes anticipatory plans, designs, and other efforts for the improvement of the urban quality of life and its built environmental architecture in the future. All these are experienced as livability and sustainability of urbanism. In the context of urban planning and policy, the improvement comprises the essential components of proximity, diversity, connectivity, and inclusivity.

Urban Intentionality, Livability and Sustainability

Urban intentionality is made possible by urbanism that is experienced by people in pedestrian public realms as a vibrant well-being of livability and sustainability. Politically speaking, livability and sustainability are demonstrated by urban citizens in a high number of their participation during the political process of public elections for mayor and city councilors; this involvement includes rallies and gatherings of the crowd in public places. Such vibrant crowds are also experienced as ecstatic jolts and joyful marches during civic holidays, periodical festivals and rituals. Most notable outdoor cultural events in the City of Toronto happen during the summer seasons, such as Canada Day Weekend, various performances of harborfront activities (Harborfront Center 2015), Dundas Square and Street festival, Danforth street festival, Toronto Caribbean carnival, Heritage Ontario festival, Beeches and Toronto Jazz festival. There are several indoor festivals as well that is participated by international organizations and persons, such as for movie, folklore, photography, storytelling and literary works. Attractions and cultural performances in public realms have been ritually celebrated by Toronto citizens and guests as part of their sense of belonging to the city they love and care for. Multicultural performing arts, foods, and drinks have been an inseparable part of the festivals. Vibrationally speaking, the urban ambiance of Toronto's public realm during the summer seasons is mostly multicultural.

From planning and policy's perspective, urban development should be in compliance with livable and sustainable principles. Regarding livability and sustainability, the City of Toronto established and developed the set of green standards such as Toronto Green Standards for new low-rise residential development (City of Toronto 2017). Accordingly, sustainability becomes clearly practical, measurable, and understandable in terms of air, water, and land quality for the healthy pedestrian environment. The implementation of the standard is for 'urban-heat-island' reduction, minimum energy performance, optimization of renewable energy, construction activity, stormwater retention and runoff, water use efficiency, urban forest, tree protection, 'canopy-tree' increase, protect-restore-enhance of natural heritage site, bird collision and mortality deterrence, reduce nighttime glare and light trespass, storage and collection of recycle organic waste, waste management for hazardous materials, and the use of local building materials.

Regarding procurement of sustainability from the municipality, new development proposals for low-rise, and mid-high-rise residential should be submitted with filled out standard checklist and statistics consisting of mandatory (Tier 1) and voluntary (Tier 2) requirements (City of Toronto 2017). These standards are designed to be in alignment with the regular development approvals and inspections process. The filled-out form of the checklist is verified on plans and drawings as well as in the reports provided as part of the regular planning application submission. Documentation requirements are set out in a development guide. Then, the green checklist, statistics template, plans and reports work together to file compliance and are reviewed by the municipal departments during the circulation process. Then, these documents are assessed by relevant municipal divisions prior to issuing planning approvals.

The urban development policy in the City of Toronto is focused on livability and sustainability. Several projects and programs are within the context of urban regeneration, revitalization, and redevelopment. Most of these are characterized by land use intensification and mixed uses. One notable example of the projects takes place in the 17 square kilometers downtown area. The intensification is worked out and achieved through the redevelopment of large brownfield areas, infill high-rise residential, and regeneration of former industrial districts into a dynamic mix of housing and workplace for inventive industries with more than 446.000 jobs and 200.000 populations in 2012; the outcome of this intensification is shown with 41% of downtown populations are walk or ride bike to work while 34% use public transit (Ostler 2014, 8). Toronto successful experience of livability and sustainability is not out of the commission of managing growth that draws people to live in an urban core area with safe, healthy, and attractive built environment.

In dealing with the urban development, the City Council has preferred to carry on services and perform certain activities through agencies and corporations. In doing so, the city agencies are independently able to meet legal requirements and work in a commercial world with a focus on delivering certain policy objective or service. Each agency and corporation has a diverse mandate and responsibility with the engagement of citizens in its board members. This participation is designed to bring specific expertise to the agency or corporation as well as to involve funders or fundraisers on a voluntary basis (Fig. 11.1).

Regarding livability and sustainability, in October 2008, Toronto City Council approved the establishment of Build Toronto Inc. under the authority of the City of Toronto Act, 2006 and Ontario Regulation 609/06 (Toronto, 1998–2017); this organization is incorporated under the Business Corporations Act, with the City of Toronto as its single shareholder. The task of the corporation is to divulge the resources in underutilized lands and use the available land base of the city and its agencies to attract targeted industries, stimulate the making of necessary employment, and revive neighborhoods. The management of the business and affairs of Build Toronto is under the supervision of the board of directors comprising 8 citizens of 11 members with the major as the chair.

For specific areas of concern, the city has similar corporations such as Build of Toronto for real estate and development, Casa Loma for a financially self-sufficient heritage attraction and hospitality venue, Invest Toronto for private engagement of investment and marketing that create

Fig. 11.1 Streetscape of Queen Street West, Toronto 2016 (Photograph by author)

desirable jobs, The Lakeshore Arena Corporation for management and the leasehold interest in the Lakeshore Arena facility, Toronto Community Housing Corporation for management of rental housing units, Toronto Hydro Corporation for electricity in Toronto, and the City of Toronto Economic Development Corporation for management of land in Ports Lands. Besides corporations, there are service agencies, such as Exhibition Place; Heritage Toronto; Civic theaters including Sony Centre for the Performing Arts, St. Lawrence Centre for the Arts and Toronto Centre for the Arts; Toronto Atmospheric Fund; Toronto Library; Toronto Parking Authority; Toronto Police Service; Toronto Public Health; Toronto Transit Commission; Toronto Zoo; and Yonge-Dundas Square.

As a whole, the governmentality of the city for planning and policy is not only supported by corporations and agencies, but it is also provided with adjudicative bodies. The bodies include Committee of Adjustment; Committee of Revision; Compliance Audit Committee; Property Standards Committee/Fence Viewers; Rooming Housing Licensing Commissioner and Deputy Commissioner; Sign Variance Committee; and Toronto Licensing Tribunal. These organizations operate independently from the city; they hold hearings to regulate activities, resolve disagreements, arbitrate on matters and determine legal rights and assistance.

Urban Intentionality and Proximity

Ontologically speaking, proximity is the spatiotemporal characteristics of being in the world (Heidegger 1927/2010, 102–5). Proximity or nearness is the phenomenon of urban intentionality that brings about people, things, and other beings closer and at the reach of hand in terms of less than 5 minutes walking distance or public transit ride from one's home place. This proximity is obvious as the basic need of a place to stay in relation to the daily consumption of foods and goods. The urban intentionality for proximity is the natural necessity for an efficient system of productions and services. In the practical world, the realization of proximity is implemented in the plan and design of density for populations and buildings in urban areas. The highly concentrated people and buildings are necessarily planned and designed to intensify activities of production, service, and leisure in such a way so that uphold, develop, and improve urban livability and sustainability.

In Toronto, urban planning and policy for proximity are understood under the notion of land use intensification and in the framework of zoning bylaw for land use, use of buildings and structures, size, density, parking and loading spaces. In order to response the dynamics of market and investment, as well as the socioeconomic transformations, zoning plans in the various area are periodically reviewed for rezoning, amendments, and adjustments. Studies and planning initiatives have been an integrated part of the rezoning process in the context of urban regenerations, place makings, infrastructure improvements, high-density lifestyle adaptation, revitalization of declining areas, and transit-oriented development. The official plan of the city sets out the urban structure and develops principles of growth management for proximity within the structure, as well as sets out policies for management of transformation of proximity through the integration of land use and transportation. In order to realize the projects, the city provides a development guide that is designed for property owners, developers, builders, and others interested in obtaining approvals for developing property in Toronto.

Livability of urbanism is not perfect without attractive built environment. On the other hand, intensification of land uses and buildings is not the best assurance for successful urban livability without permeability, connectivity, and visually appealing form and space of a safe pedestrian environment. Attention to the significance of urban design has been an integrated part of urban development in Toronto since 2000; the momentum for this was when the City of Toronto established urban design award in 2000 and urban design panel in 2005.

Urban design review panel is an independent advisory body concerning urban design policy. Their advice is based on professional judgment, understanding of principles for design excellence, in compliance with the official plan and other related documents, such as design guidelines, secondary plans, and the design quality of the designated project. Their members consist of some of the leading architects, landscape designers, engineers, and planners. The mandate of this panel is to guide and inform the architectural and technical qualities of design for all projects proposed in the designated area. The panel's task is to set a comprehensive design standards excellence so that the designated areas own unique architectural identity and vibrant public realm. The panel provides design advice and considerations to city staffs that enable them to improve the quality of design concerning the public realm. The scope of the panel's work includes both public works—such as parks, bridges, and other infrastructures—and private projects, including residential, commercial, and retail buildings. The first project that had initiated and involved the design panel was the Toronto Water Front Development in 2005. Today, the design panel review is not only to concern with new developments, but also to engage with most design matters including preserving the distinctiveness of place, maintaining liveliness, ensuring comfort and public safety, and harmonizing new development with its contextual surroundings.

Regarding density, the planning and policy of the City of Toronto are focused on the land use policy for growth that regards the existing physical character and its resources (City of Toronto 2015, Chapter 4, 103–6). The management of growth is to integrate broad range employment areas—such as suburban office parks—and institutional areas—hospitals, college, and university campuses—with public transit that reinforce the existing physical character and its environmental resources. In the context of the neighborhood, the policy of density is focused on harmonizing new developments with the existing architectural character and environmental conditions of designated locations that maintain and sustain their social and ecological stability. In the context of infill development, several best practices have been done successfully in the past decades that by design, the architecturally harmonious integration of high-density residential towers with the architecturally established neighborhood with historical character is possible (Fader 2000). This harmonization policy includes massing and the design treatment for the transition between high and low as well as large and small form. Nevertheless, planning and policy of density entail the creation of livable streetscapes and pedestrians with the support of small retails and commercial services. The sense of density by design is shown by the integration of neighborhoods, employment, and institutional areas into the public transit network.

Urban Intentionality and Diversity

Diversity matters in business society because it provides competitive differences among agencies that create, develop, and sustain productive opportunities and platforms for the best in terms of processes, systems, and procedures (Partridge 1999, 91; Cornelius 2001, 60–61; Pride 1992, 266). In the natural sciences, biodiversity has been scientifically proven as the necessity for better evolution, agriculture, domestication, habitation, and environmental sustainability of all species on earth (Wilson and Peter 1988; Gepts 2012; Schulze and Mooney 2012; MacArthur and Wilson 1967). Accordingly, diversity is regarded as the necessary condition for the ecological processes. However, diversity is imperatively integrated within the system of production that maintains, develops, and sustains within the carrying capacity of the system so that the harvesting rates are not over the regeneration rates. This equilibrium applies as well for emissions and assimilative capacity of natural, economic, and sociocultural productions.

For urbanism, the concept of diversity has been widely recognized as the essential condition for land use planning that matters for the urban economy and its regional growth. Historically speaking, this recognition is regarded as a critical response to the early twentieth-century modern movement of architecture and urbanism. The modern architects and city planners were under attack for the decline of vibrant urbanity in various cities in North America because of their strictly programmatic division of land use into work, live, and play (Jacobs 1961). Since the 1970s, the necessity for mixed land use and populations becomes the essential need of urbanism under various programs of urban development such as revitalization, rejuvenation, gentrification, and urban renewal. Accordingly, the question of diversity remains important for the relationship between populations and vibrant urbanism for the healthy and sustainable community.

In the City of Toronto, diversity is characterized by its demography, land use, architecture, transportation, and landscape. Demographically speaking, there are more than 140 languages spoken in the city. Based on an empirical study of Vipond (*Making a Global City: How One Toronto School Embraced Diversity*, 2017), multicultural citizenship is a synergy of the past legacy of memories and the present-day consent toward an undivided society. Accordingly, embracing newcomers from all over the globe has been empirically demonstrated as an enrichment and enhancement of historically established culture. Diversity is not a choice but a necessity for growth economically, politically, socially, emotionally, and spiritually. School is one of the best platforms and environments for exercising diversity as an integrated part of the learning process toward a multicultural society. The other environment that entails diversity is workplace; the integration of diversity into workforce plan for the city is incorporated within the 2015–2018 Strategic Plan (City's Manager Office_City of Toronto 2015). In urban planning and policy, the necessity for the diversity encourages various studies and initiatives for urban regeneration projects with mixed uses development.

Several case studies confirm the contribution of mixed uses development for livability and sustainability of urbanism (Coupland 1997). Unsurprisingly, to understand why the Urban Land Institute consolidated theories and best practices of mixed-use development into a handbook (Urban Land Institute 2003). Urban growth in terms of employment, economy, and culture is necessarily provided by diverse populations, activities, and functions that establish and intensify interactions and learning processes toward new things and events. Festivals and rituals of the city are upheld and enriched by diverse communities and people. In order to support urban growth, planning and policy for diversity are necessarily synchronized with urban design policy in shaping and encouraging safe and attractive public places. The sense of urbanism is experienced as the ecstatic self-actualization of the crowd in celebrating individual freedom in anonymity.

In neighborhood context, plural community—in terms of ages, incomes, social milieu, background, occupation, sexual orientation, and socio-religious affiliation—is theoretically more social intelligence, open-minded, and tolerance compared to those who live in relatively singular group identity. For the City of Toronto, the culturally plural world is experienced on daily basis in various realms such as schools, workplaces, public transits, and public places. Culturally speaking, with respect to differences in the context of multiculturalism has been a historically integrated part of Canadian identity (Mackey 1999). This identity is demonstrated by Canadians in their attitude toward public policy that is not simply about a universal healthcare system. Rather this includes their attitude to gun control, abortion rights, gay marriage, decriminalization of marijuana, and non-Islamic phobia (Tomalty and Mallach 2016, Chapter 2).

Urban Intentionality and Connectivity

Human existence is characterized by connectivity. This connectivity manifests in manifold forms, ways, manners, and styles because by nature, all human minds are subconsciously connected and able to understand each other non-verbally (Neuliep 2009, 247; Guerrero and Farenelli 2009; Hinde 1972). In the urban context, connectivity manifests in the integration of all places into a network of spatial ecosystem and the infrastructure for communications, transportations, and utilities. In doing so, there is no place in the urban area left behind and inaccessible and unreachable by any means. Connectivity is vital and essential for the livability of urbanism in relation to economic growth (Chakwizira 2015, 212), social inclusivity and sustainability (Manzi et al. 2010), and ecological integrity (Sagoff 1995).

Urban planning and policy are necessary to ensure that all areas and points of destinations are accessible and reachable with public transit and utilities. As a whole, the sense of connectivity is to bring people together in terms of ideas, efforts, funds, talents, and skills that enable them to understand each other and work together. Physically speaking, urban connectivity concerns a technically integrated system of infrastructure for communication, transportation, and utilities. All this is necessary reliable to support urban activities in terms of productions, services, and leisure. In order to ensure the connectivity of urbanism, the City of Toronto established and developed infrastructure policy and standards that apply to the public and private property.

The City of Toronto is one of the best examples for a well-integrated system of land use and transit network plan. The use of public transit has been growing in this city because the city has established and developed a deliberate policy of transit-oriented development for decades; between 1960 and 1980, the use grew 48% (Newman 1996, 86–87). The public transit system in this city is not only safe and reliable but also convenient with several designated waiting areas and junctions for buses, subways, and streetcars. To a certain extent, the success of Toronto's public transit policy is partly an outcome of federal tax laws that discourage people to own single-family houses; most residents live in high-density multifamily housing and apartments because there is less incentive for mortgage interests and property taxes (Cervero 1998, 86). In a matter of fact, Toronto in 2017 is the North American urban center with a population density of 4149.5 people per square kilometer or 10,750 in a square mill (World Population Review 2017).

Urban Intentionality and Inclusivity

Inclusivity is an essential principle of livability and sustainability that bring about people and other resources within its urban domain as a whole. Citizen participation is one aspect of urban democratic inclusivity. In the City of Toronto, urban inclusivity is designed to engage citizens in the planning of the urban development process. The Toronto planning review panel is an innovative way for residents to become involved in city planning processes. The panel began in September 2015; the 28 members of the panel were selected through a random selection process known as a civic lottery. Accordingly, 12,000 randomly nominated households in Toronto received invitations to volunteer to be a member of the panel for a two-year term. This process is to ensure that the members represent the diverse populations of Toronto while broadening the appointment by bringing new voices into the democratic planning process of the city.

In the broadest sense of inclusivity, the City of Toronto delivers and provides several opportunities for its citizens to get involved in various programs and activities (City of Toronto 2017). This engagement includes voting for local elections, joining an advisory group—such as for community planning, infrastructure improvement, and natural heritage conservation—taking part in public consultations, having your say through city councilor, taking part in participatory budgeting, and participation in the city councilor's meetings as an observer. All this is intentionally for the notion and practice of open governmentality and public accountability. In doing so, the management and operation of the livability and sustainability of urbanism are publicly transparent.

Intentionality and Global Urbanism

In the Age of Global Information and economy, proximity and connectivity are understood in the framework of within the cyber-connection. Distance becomes mere seconds to click or to hit the name or sign on the touch screen of the wireless communication device. Urban connectivity in this cyber realm is not confined by political geography or physical domain, but by personal preference and choice for whom are in his/her list of contacts. The contacts represent members of a community or a group which is designated by the owner of the device according to his/her categorical favorites. The public realms of cyber community manifest as various social media such as Twitter and Facebook; people share publicly ideas, interests, opinions, and information through technologically computing media with interactive capability on the cyber network of connectivity. On this connectivity, being an acolyte of social media has become a precondition for the construction of a personal identity (James and Steger 2016, 21). The transformation of urbanism happens through the way people communicate and interact with each other. This transformation happens dramatically to challenge several concepts and principles of urbanism from its intentionality to practicality.

Intentionally speaking, the global information provides urbanism with the global consciousness based on personal choice and preference. Theoretically speaking, there is the necessity for redefining the reality of urbanism from this cyber connectivity. To what extent is this virtual connectivity constituent to the concept and practice of urbanism? In terms of concept, urbanism can be defined as a system of ideas, efforts, and activities for making a place to work, stay, and play with highly concentrated populations and buildings with a sociopolitically organized body. Connectivity plays a significant role in this urban world that enables them to organize, mobilize, produce, and distribute resources through the use of language and technological means. Thus, connectivity is about communication and interaction.

Cyber connectivity is distinguished from the physical one because of its method to convey content in terms of practicality. Based on this virtual conductivity, the practice of urbanism changes from physically based communication to virtually based interaction. However, with this transformation, the concept of urbanism is challenged to expand its conceptual boundary but it does not need to drop its intentionality: the human decency of social being or urbanity. In a matter of fact, the conceptual boundary of urbanism is made possible by the livability and sustainability of the urban community. In other words, cyber connectivity reinforces and enhances the presence and existence of an urban community in the age of technologically global information with cyber connectivity. The notion of public accountability becomes more effective and broader in its accessibility and transparency because people are able to observe what is going on in their city from the city's Web site on their personal wireless device. Nevertheless, cyber connectivity challenges urbanism concerning the concept of inclusivity that is not simply from the geographical and normative definition of citizenship. The cyber connectivity is able to draw diaspora, who is originally from the city, from anywhere on the globe to vote or to participate in urban development.

Indeed, technologically global connectivity has been playing a significant role in the formation, development, and sustainability of urbanism in almost everywhere on the globe since the first millennium. However, the global capitalism that has been shaping and dominating the cities on the globe since the eighteenth century through the trade of spices, tea, tobacco, coffee, and slaves. All this have been mostly originated from a long histor-ical development of European mercantilist imperialism and colonialism since the seventeenth century or earlier that established the global networks of trade and commerce until today. Urbanization in industrial countries has been involving and being resulted from the dynamics of intertwining power between capital and labor. The City of Toronto is historically insep-arable from the global capitalist expansion of British and French explor-ers and traders in North America between 1730 and 1840 (Wallerstein 2011). The American revolutionary war drew many British loyalists to set-tle in Toronto; French interest in Canada was more driven by economic opportunity than by colonial settlement (Scott 2011, 234). Economi-cally speaking, Toronto was a well-known center of alcohol distillation in North America; the Gooderham and Worts Distillery operations became the world's largest whiskey factory by the 1860s (Pateman 2013, 48).

Like other democratic urban centers in the Global North, Toronto has been managing to minimize the consequences of global capitalist hege-mony on urbanism. The threat of this hegemony is the accumulation of capital power that is able to put the welfare state and social justice in peril. So that our concept on capital is necessary to be redefined in the broadest sense of the word as cultural—a wealth of knowledge—social—a wealth of networks—and symbolic power that includes all of its resource-ful potentiality for interest-oriented actions (Bourdieu and Thompson 1991; Bourdieu 1986). Accordingly, the notion of capital should be not understood merely as a financial power and economic resource. Rather, capital should include political, social, and cultural power and resources that are able to converse as symbolic power. Since the various forms of capital: Social, cultural, political, and economical are intertwined together within the practices of governing (Olson 2008).

In dealing with the global economy, the City of Toronto chooses to face it with Toronto's global competitiveness. This competitive edge is based on the facts that Toronto stands out among other global cities in North America because it provides with 1.4 million highly skilled and multilingual workforce, 89,800 diverse businesses, internationally high-ranked post-secondary educational facilities, 64% populations between 25 and 65 age with university degree, economically proven resilience to global economic downturn, technological sector with diverse 14.600 companies and about 159.000 people, financial institutions with the international's soundest for six years after World Economic Forum, and international's most competitive major city (City of Toronto 2017).

Concluding Remarks

What does make people together for urbanism? Why are people attracted to work, live, and play together in the urban realm? They do not each other and come from various backgrounds and origins, but they feel and share something in common about the mentally driving necessity for being a member of the urban community. In such a community, they are able to find their own way to work, live, and play that grows together. Ben Anderson speaks of imagined communities (Anderson 2006). Phenomenological philosophy after Husserl (Husserl 1962/2002/2014) recognizes this mentally driving force as intentionality. The questions mentioned above lead us to inquiry human consciousness that drives and brings their ideas, efforts, strives, endeavors, and feelings toward urbanism; this is the phenomenon of gathering, which is characterized with highly concentrated populations and buildings in a location.

Until recently, the relationship between urbanism and consciousness has been fallen out of attention in urban studies for many reasons. The importance of this relationship has been argued by German urban sociologist Georg Simmel in his 1903 essay: The Metropolis and the Mental Life (Simmel 1950) and by Walter Benjamin in his Berlin Chronicle (Benjamin, *A Berlin Chronicle Notices*, 2015) & *Berlin Childhood Around 1900* (Benjamin, *Berlin Childhood Around 1900* 2006). Simmel and Benjamin share something in common concerning urbanism that is characteristically signified with a phenomenon of the anonymous crowd; this phenomenon is experienced as a reality of collectively united intention but with socially dispersing individuality. As a matter of fact, the spiritually unifying power of urbanism is not consciously organized as a sociopolitical movement. Rather, this power works and operates at the subconscious level of people but it brings people together as an urban community. The existence of this power is experienced as urban chore or routine that manifests in various forms, ways, styles, and manners of working, living, and recreation.

Urban conscious awareness is the mental states or events that enable people to understand and engage into the urban life-world. Intentionally speaking, urban people are those who spiritually driven to gather things and others into their nearness and to stand out collectively for human decency. In the City of Toronto, urbanism has been demonstrated to get into urban livability within the framework of socioeconomic and environmental sustainability. Planning and policy are directed and managed to achieve this goal through the principles of the proximity of settlement, a well-integrated system of transit and land use, and democratic inclusivity of its citizens.

The consciousness of global urbanism is technologically made possible by the advent of information technology. The essential structure of this consciousness is the global-wide connectivity of people through technologically mediated communication with Internet access and network. The connectivity leads us to go deeper into the essential need of humankind as social being for proximity or nearness. This is actually also the intentionality of urbanism at a spiritual level. Sociologically speaking, urban community is a complex system consisting of diverse members and various structures that work and operate for different goals for urban livability and sustainability.

However, they are united for one ultimate objective: human decency; this is what we call urbanity. Then, the question of urbanism is the problem of humankind as a social being that they are by nature is driven to work together for their presence and existence with dignity. Urban livability is about the state of vibrant well-being for urban settlement. This vibration is experienced by people as the capricious sphere of safe, healthy, and lively casual crowd and streams in public realms on pedestrian environments and public transits. While urban sustainability is the quality of the urban settlement system that is to manage their diverse resources for various productions, services, and recreations indefinitely, urban livability is the system of happenings when the urban potentials come into play with vibrant activities. The sense of livability and sustainability of urban phenomenon lies in its care and capacity to uphold and maintain human dignity and environmental conservation.

References

Amstrong, Frederick H. 1988. *A City in the Making: Progress, People and Perils in Victorian Toronto*. Toronto and Oxford: Dundurn.

Anderson, Benedict. 2006. *Imagined Communities: Reflection on the Origin and Spread of Nationalism*. London and New York : Verso.

Benjamin, Walter. 1999. *The Arcades Project*. Edited by Rolf Tiedemann. Translated by Howard Eiland and Kevin McLaughlin. Cambridge, MA: Belknap Harvard University Press.

———. 2006. *Berlin Childhood Around 1900*. Cambridge, MA: Harvard University Press.

———. 2015. *A Berlin Chronicle Notices*. Translated by Carl Skoggard. New York: Publication Studio.

Berger, Hans. 1929. Ueber das Elektronenkephalogram des Menschen. *Arch-Psychiatr Nervenkrankenheit* 87: 527–570.

Bourdieu, Pierre. 1986. "The Form of Capital." In *Handbook of Theory and Research for the Sociology of Education*, edited by J. Richardson, translated by R. Nice, 241–58. New York: Greenwood.

Bourdieu, Pierre, and John B. Thompson. 1991. *Language and Symbolic Power*. Cambridge, MA: Harvard University Press.

Buzsaki, Gyorgy. 2006. *Rhythms of the Brain*. Oxford: Oxford University Press.

Cervero, Robert. 1998. *The Transit Metropolis: A Global Inquiry*. Washington, DC: Island Press.

Chakwizira, James. 2015. "The Urban and Regional Economy Directing Land Use and Transportation Planning and Development." In *Land Use Management and Transportation Planning*, edited by C. B. Schoeman, 199–225. Hampton and Boston: WIT Press.

City of Toronto. 2015. "Toronto Official Plan." Accessed August 16, 2017. https://www1.toronto.ca/wps/portal/contentonly?vgnextoid=03eda07 443f36410VgnVCM10000071d60f89RCRD.

———. 2017. "Toronto Green Standard: Making a Sustainable City Happen." Accessed August 17, 2017. https://www1.toronto.ca/City%20Of% 20Toronto/City%20Planning/Developing%20Toronto/Files/pdf/TGS/2017 TGS_LowRise_Standard.pdf.

City's Manager Office_City of Toronto. 2015. "2015–2018 Strategic Plan." Accessed August 14, 2017. https://www1.toronto.ca/City%20Of% 20Toronto/Equity,%20Diversity%20and%20Human%20Rights/Divisional %20Profile/Policies%20-%20Reports/A1503399_Strat_Plan_web.pdf.

Cornelius, Nelarine. 2001. *Human Resource Management: A Managerial Perspective.* London: Thompson.

Coupland, Andy. 1997. *Reclaiming the City: Mixed Use Development.* London: Spon.

Fader, Steven. 2000. *Density by Design: New Directions in Residential Development.* Washington, DC: Urban Land Institute.

Gepts, Paul. 2012. *Biodiversity in Agriculture: Domestication, Evolution, and Sustainability.* Cambridge UK: Cambridge University Press.

Goldberg-Miller, Shoshanah B. D. 2017. *Planning for a City of Culture: Creative Urbanism in Toronto and New York.* New York and London: Routledge/Taylor & Francis.

Guerrero, Laura K., and Lisa Farenelli. 2009. "The Interval of Verbal and Nonverbal Codes." In *21st Century Communication: A Reference Handbook,* edited by William F. Eadie, vol. 1, 239–48. London: Sage.

Harborfront Center. 2015. "Harborfront Center." Accessed August 17, 2017. http://www.harbourfrontcentre.com/.

Heidegger, Martin. 1927/2010. *Being and Time.* Edited by Dennis J. Schmidt. Translated by Joan Stambaugh. Albany, NY: SUNY Press.

Hinde, Robert A. 1972. *Non-verbal Communication.* Cambridge: Cambridge University Press.

Husserl, Edmund. 1901–1902/2001. *Logical Investigations,* vol. 1. Translated by N. Findlay. London: Routledge.

———. 1962/2002/2014. *Ideas, General Introduction to Pure Phenomenology.* Translated by W. Boyer Gibson. London and New York: Routledge.

Jacobs, Jane. 1961. *The Death and Life of the Great American Cities.* New York: Random House.

James, Paul, and Manfred B. Steger. 2016. "Globalization and Global Consciousness: Levels of Connectivity." In *Global Culture: Consciousness and Connectivity,* edited by Roland Robertson and Didem Buhari-Gulmez, 21–40. Abingdon and New York: Ashgate/Routledge.

Lemon, James T. 2008. *Liberal Dreams and Nature's Limits: Great Cities of North America Since 1600.* Eugene, OR: Wipf & Stock.

Levine, Allan. 2014. *Toronto: Biography of a City.* Madeira Park, BC: Douglas & McIntyre.

MacArthur, Robert Halmer, and Edward O. Wilson. 1967. *The Theory of Island Biogeography.* Princeton, NJ: Princeton University Press.

Mackey, Eva. 1999. *House of Difference: Cultural Politics and National Identity in Canada.* London: Routledge.

Mackintosh, Phillip Gordon. 2017. *Newspaper City: Toronto's Street Surfaces and the Liberal Press, 1860–1935.* Toronto: University of Toronto Press.

Manzi, Tony, Karen Lucas, Tony Lloyd Jones, and Judith Allen. 2010. *Social Sustainability in Urban Areas: Communities, Connectivity and the Urban Fabric.* London: Routledge.

Neuliep, James William. 2009. *Intercultural Communication: A Contextual Approach.* London: Sage.

Newman, Peter. 1996. "Reducing Automobile Dependence." *Environment and Urbanization* 8 (1): 67–72.

Olson, Kevin. 2008. "Governmental Rationality and Popular Sovereignty." In *No Social Science Without Critical Theory*, edited by Harry F. Dahms, vol. 25, 329–52. Bingley: Emerald.

Ostler, Thomas. 2014. *Downtown Toronto: Trends Issues Intensification.* Toronto, ON: City of Toronto, City Planning - Toronto and East York District.

Pateman, John. 2013. *Fort Pitt to Fort William.* Thunder Bay: Pateran Press.

Partridge, Lesley. 1999. "Creating Competitive Advantage with HRM." www.selectknowledge.com. Select Knowledge.

Pride, William M. 1992. *Business.* Boston: Houghton Mifflin.

Redway, Alan. 2014. *Governing Toronto: Bringing Back the City That Worked.* Victoria: Friesen.

Sagoff, Mark. 1995. "The Value of Integrity." In *Perspectives on Ecological Integrity*, edited by Laura Westra and John Lemons, 162–76. Dordrecht: Springer Science & Business Media.

Schulze, Ernst-Detlef, and Harold A. Mooney. 2012. *Biodiversity and Ecosystem Function.* Berlin, Heidelberg, and New York: Springer Science & Business Media.

Scott, Bruce R. 2011. *Capitalism: Its Origins and Evolution as a System of Governance.* New York, Dordrecht, and Heidelberg: Springer Business Media.

Searle, John R. 1983. *Intentionality: An Essay in the Philosophy of Mind.* Cambridge: Cambridge University Press.

Sewell, John. 1993. *The Shape of the City: Toronto Struggles with Modern Planning.* Toronto, Buffalo, and London: University of Toronto Press.

Simmel, Georg. 1950. "The Metropolis and Mental Life." In *The Sociology of Georg Simmel*, edited by D. Weinstein, translated by Kurt Wolff, 409–24. New York: Free Press.

Stanwick, Sean, Jennifer Flores, and Tom Arban. 2007. *Design City Toronto.* Toronto: Wiley.

Statistics Canada. 2009-11-20. "Immigration in Canada: A Portrait of the Foreign-Born Population, 2006 Census: Portraits of Major Metropolitan

Centres." Accessed August 9, 2017. http://www12.statcan.ca/census-recensement/2006/as-sa/97-557/p24-eng.cfm.

The City of Toronto, Ontario Canada. 2002. "Toronto Official Plan." Accessed August 10, 2017. https://www1.toronto.ca/planning/chapters1-5.pdf#page=17.

Tomalty, Ray, and Alan Mallach. 2016. *America's Urban Future: Lessons from North of the Border.* Washington, DC: Island Press.

Tonino, Guilio. 2003. "Consciousness Differentiated and Integrated." In *The Unity of Consciousness,* edited by Axel Cleeremans, 253–65. Oxford: Oxford University Press.

Urban Land Institute. 2003. *Mixed-Use Development Handbook.* Washington, DC: Urban Land Institute.

Vipond, Robert. 2017. *Making a Global City: How One Toronto School Embraced Diversity.* Toronto, Buffalo, and London: University of Toronto Press.

Wallerstein, Immanuel. 2011. *The Modern World-System III: The Second Era of Great Expansion of the Capitalist World-Economy, 1730s–1840s.* Berkeley, Los Angeles, and London: University of California Press.

Wilson, Edward O., and Frances M. Peter. 1988. *Biodiversity.* Washington, DC: National Academy Press.

World Population Review. 2017. "Toronto Population 2017." Accessed August 19, 2017. http://worldpopulationreview.com/world-cities/toronto-population/.

Postscript

The phenomenon of urbanism is essentially conceived and established by human condition as a social being, which is characterized by fear of isolation and alienation and by a passion for connection. This fear is intrinsic and perpetual that potentially drags people down emotionally and spiritually into alienation, despair, and desperation. On the other side, the human condition is passionate in the search for proximity. Proximity for humankind is more than just the necessity for communication and interaction that enable them to understand each other, but also the physical and social nearness that enable them to share, develop, and sustain the world as their home. Hence, this proximity is the existential imperative for organization and cooperation. In order to work effectively and efficiently, diversity plays an important role so that people are able to find their own niche of dexterity even in an anonymous crowd. Thus, proximity for humankind is not only the state of affairs of nearness, but it needs diversity and dexterity so that they are able to grow and cultivate their biological, social, and spiritual needs.

The existential imperative for proximity is possible because humankind is innately endowed with language competence. In this sense, language competence is not only the ability of communication and interaction, but

B. Wiryomartono, *Livability and Sustainability of Urbanism,*
https://doi.org/10.1007/978-981-13-8972-6

the innate capacity of identification, organization, and construction of the self and others in various ways and fashions for services and productions. Based on this capacity, humankind is able to develop, utilize, and mobilize their resources to build a collective home for growing together as a sociopolitical and economic community. Urbanism is a strategic endeavor of such a community for having a home on the earth under the sky. This is nothing but to prevent, deal with, and overcome their existential anxiety of nothingness in terms of seclusion and alienation. The passion for proximity leads humankind to explore and develop highly sophisticated styles, ways, fashions, manners of their relationship with others to overcome their fear of mortality. Monumental structures and places have been conceived as the embodiment of unbreakable humanity against climate and time. All of these are conceptually integrated into a concept we call urbanity. Urbanity is considered as the utmost endeavor of humankind as social being for urbanism.

Urbanism is a cultural endeavor of people who work, live, and play together because they share and have something in common for their life-world in a highly concentrated settlement for productions, services, and recreations. Urbanism is a complex and highly sophisticated effort because it involves various skills, resources, and disciplines. Theoretically speaking, the chapters of this book have explored and unfolded the aspects, elements, and structures history, philosophy, planning, design, economy, geography, arts, architecture, political sciences, and ecology that contribute to the processes, constructions, and operations of the urban life-world. The necessity for the safe, healthy, and attractive built environment is not the work overnight and the one-man-show achievement but the incremental outcome of evolutionary development with trials and errors with collaborations and confrontations of urban inhabitants.

Despite the complexity and multidimensionality characterize the urban life-world, urban planning and design are considered the best rational and artistic attempts to manage of urban development and conservation that work for the economic and sociocultural growth of the city. Plans and designs are necessary open-ended that provide the frameworks, guidelines, and directions of urban development and conservation. Good plans and designs for urbanism should involve all people according to their specific role, function, responsibility, and contribution so that the urban life-world is an inclusive economic and sociocultural community.

Nevertheless, the actualization of plans and designs for urbanism is the daily experience of people in their public realms and streets. The quality of urbanism is about the manners and comportments of people in their social interactions in public spaces either on the feet or on the wheels. The examples from Malaysia and Toronto Canada demonstrate that urbanism is a cultural endeavor with local and global appearances, representations, and contents. Urbanism is always contextual because of its historical processes and its geographical settings. Regardless of these contextualities, the essence of urbanism is the search for a home in the anonymous crowd with a steady effort for composure from various contradictions, transformations, fluctuations, and uncertainties. Climate changes and global capitalist economy are the challenges of urbanism that endanger the sense of home from the environmental and existential dimension. Historically speaking, there are no challenges without the way out, solution, and resolution because urbanism is a collective effort and endeavor for beings as a whole. Crises and disasters are not only the wake-up calls for the necessity for the wholeness of the urban life-world but also the processes and passages of recognition and cultivation for being more human with ecologically rounded awareness.

References

Abdullah, Asma. 1996. *Going Local: Cultural Dimensions in Malaysian Management*. Kuala Lumpur: Malaysian Institute of Management.

Abrahamson, Mark. 2004. *Global Cities*. Oxford: Oxford University Press.

Acuto, Michele. 2013. *Global Cities, Governance and Diplomacy: The Urban Link*. London and New York: Routledge.

Adams, Kathleen, and Kathleen Gillogy. 2011. *Everyday Life in Southeast Asia*. Bloomington, IN: Indiana University Press.

Adler, Philip J., and Randall L. Pouwels. 2007. *World Civilizations, Since 1500*. Boston: Cengage.

Alhady, Alwi. 1965. *Adab Tertib*. Kuala Lumpur: Malay Publication.

Alexander, Christopher. 1979. *The Timeless Way of Building*. Oxford: Oxford University Press.

———. 1987. *A New Theory of Urban Design*. Oxford: Oxford University Press.

Amen, Michael Mark, Kevin Archer, and Martin M. Bosman. 2006. *Relocating Global Cities: From the Center to the Margins*. Lanham: Rowman & Littlefield.

American Planning Association. 2006. *Planning and Urban Design Standards*. Hoboken, NJ: Wiley.

Amstrong, Frederick H. 1988. *A City in the Making: Progress, People and Perils in Victorian Toronto*. Toronto and Oxford: Dundurn.

© The Editor(s) (if applicable) and The Author(s), under exclusive license to Springer Nature Singapore Pte Ltd. 2020
B. Wiryomartono, *Livability and Sustainability of Urbanism*,
https://doi.org/10.1007/978-981-13-8972-6

Andaya, Barbara Watson, and Leonard Y. Andaya. 1982. *A History of Malaya*. Houndmills: Macmillan.

Anderson, Benedict. 2006. *Imagined Communities: Reflection on the Origin and Spread of Nationalism*. London and New York: Verso.

Anderson, Benedict, James Siegel, and Audrey Kahin. 2003. *Southeast Asia Three Generations: Essays Presented to Benedict Anderson*. Ithaca, NY: SEAP Publications.

Andranovich, Gregory, and Gerry Riposa. 1993. *Doing Urban Research*. New York: Sage.

Ang, Ien. 2001. *On Not Speaking Chinese: Living Between Asia and the West*. London and New York: Routledge.

Arendt, Hannah. 1958/2013. *Human Condition*. Chicago: University of Chicago Press.

Aristotle. 1983. *Selected Works*. Edited by Hippocrates G. Apostle and Lloyd P. Gerson. Grinnell, IA: Peripatetic Press.

———. 1997. *Politics, Volume IV*. Edited by Richard Kraut. Oxford: Oxford University Press.

Aristotle, and Richard Kraut. 1997. *Politics*. Oxford: Oxford University Press.

Aristotle, and Richard McKeon. 1992. *Introduction to Aristotle*. New York: Modern Library.

Asiapac Editorial. 2003. *Gateway to Malay Culture*. Singapore: Asiapac Books.

Atkin, Tony, and Robert Rykwert. 2005. *Structure and Meaning in Human Settlement*. Philadelphia: University of Pennsylvania Museum of Archeology and Anthropology.

Azoulay, Vincent. 2014. *Pericles of Athens*. New York: Princeton University Press.

Bachelor, Peter, and David Lewis. 1986. *Urban Design in Action: The History, Theory, and Development of the American Institute of Architects' Regional/Urban Design Assistance Teams Program*. Raleigh: School of Design NC State University.

Bacon, Edmund. 1967. *Design of Cities*. London: Thames & Hudson.

Baharuddin, Syamsul Amri. 2008. "Competing Domains of Controls: Islam and Human Rights in Malaysia." In *Islam and Human Rights in Practice, Perspectives Across Ummah*, edited by Shahram Akbarzadeh and Benjamin McQueen, 1081–117. London and New York: Routledge.

Bairoch, Paul. 1991. *Cities and Economic Development: From the Dawn of History to the Present*. Translated by Christopher Braider. Chicago, IL: University of Chicago Press.

Banhart, Robert K., and Sol Steinmetz. 1988. *The Barnhart Dictionary of Etymology*. Hackensack, NJ: H. W. Wilson.

Barker, Ernest. 1948. *Traditions of Civility: Eight Essays.* Oxford: Oxford University Press.

Barker, Graeme. 1999. *Companion Encyclopedia of Archaeology.* London: Routledge.

Barker, Sir Ernest. 1958/1998. *The Politics of Aristotle.* Oxford: Oxford University Press.

Barnett, Jonathan. 1974. *Urban Design as Public Policy: Practical Methods for Improving Cities.* New York: Architectural Record Books.

BBC World Service. 2016. "www.globescan.com." April 27. Accessed April 28, 2016. http://www.globescan.com/images/images/pressreleases/BBC2016-Identity/BBC_GlobeScan_Identity_Season_Press_Release_April%2026.pdf.

Beatley, Timothy. 2005. *Native to Nowhere: Sustaining Home and Community in a Global Age.* Washington, DC: Island Press.

Beatley, Timothy, and Peter Newman. 2009. *Green Urbanism Down Under: Learning from Sustainable Communities in Australia.* Washington, DC: Island Press.

Beere, Jonathan. 2009. *Doing and Being: An Interpretation of Aristotle's Metaphysics Theta.* Oxford: Oxford University Press.

Bellina, Berenice, and Ian Glover. 2004. "The Archaeology of Early Contact with India and the Mediterranean World from the Fourth Century BC to the Fourth Century AD." In *Southeast Asia from Prehistory to History,* edited by Ian Glover and Peter Bellwood, 68–88. Oxfordshire: RoutledgeCurzon.

Benjamin, Walter. 1999. *The Arcades Project.* Edited by Rolf Tiedemann. Translated by Howard Eiland and Kevin McLaughlin. Cambridge, MA: Belknap Harvard University Press.

———. 2006. *Berlin Childhood Around 1900.* Cambridge, MA: Harvard University Press.

———. 2015. *A Berlin Chronicle Notices.* Translated by Carl Skoggard. New York: Publication Studio.

Bennet, Rudolf. 2005. "Husserl's Concept of the World." In *Edmund Husserl Critical Assessments of Leading Philosophers,* edited by Rudolf Bennet, Donn Welton, and Gina Zavota. New York: Routledge.

Berger, Hans. 1929. "Ueber das Elektronenkephalogram des Menschen." *Arch-Psychiatr Nervenkrankenheit* 87: 527–70.

Berger, Peter, and Thomas Luckmann. 2011. *The Social Construction of Reality: A Treatise in the Sociology of Knowledge.* New York: Open Road.

Berghuis, Thomas. 2012. "Architecture." In *Cultural Sociology of the Middle East, Asia and Africa: An Encyclopedia,* edited by Andrea Stanton and Edward Ramsamy, 204–9. London: Sage.

Bertrand, Jacque. 2013. *Political Change in Southeast Asia.* Cambridge: Cambridge University Press.

Binne, Jon, Julian Holloway, Steve Millington, and Craig Young. 2006. *Cosmopolitan Urbanism.* London and New York: Routledge.

Birch, Eugenie L., and Susan Wachter. 2011. *Global Urbanization.* Philadelphia: University of Pennsylvania Press.

Bishop, Ryan, John Philips, and Wei Yeo Wei. 2003. *Postcolonial Urbanism: Southeast Asian Cities and Global Processes.* London and New York: Routledge.

Blackwood, William. 1822. "Blackwood Magazine." *Edinburg Magazine,* 346.

Blau, Judith R., Mark La Gory, and John Pipkin. 1983. *Professionals and Urban Form.* Buffalo: SUNY Press.

Bogucki, Peter. 1999. *The Origins of Human Society.* London: Blackwell.

Bourdieau, Pierre, and John B. Thompson. 1991. *Language and Symbolic Power.* Cambridge, MA: Harvard University Press.

Bourdier, Jean Paul, and Nezar AlSayyad. 1989. *Dwellings, Settlements and Traditions: Cross-Cultural Perspectives.* Lanham, MD: University Press of America.

Bourdieu, Pierre. 1982. *The Logic of Practice.* Stanford: Stanford University Press.

———. 1984. *Distinction: A Social Critique of the Judgment of Taste.* London: Routledge.

———. 1986. "The Form of Capital." In *Handbook of Theory and Research for the Sociology of Education,* edited by J. Richardson, translated by R. Nice, 241–58. New York: Greenwood.

———. 1990. "Structures, Habitus, Practices." In *The Logic of Practice,* edited by P. Bourdieu, 52–79. Stanford, CA: Stanford University Press.

Bowman, Alan K., Peter Garnsey, and Dominic Rathbone. 2000. *Cambridge Ancient History,* vol. 11. Cambridge UK: Cambridge University Press.

Bowra, C. M. 1965. *Classical Greece, Great Ages of Man.* New York: Time Life Books.

Branigan, Keith. 2002. *Urbanism in the Aegean Bronze Age.* New York: Sheffield Academic Press.

Bravi, Alessadra. 2015. "The Art of Late Antiquity: A Contextual Approach." In *A Companion to Roman Art,* edited by Barbara Borg, 130–50. Malden: Wiley.

Brenner, Neil, and Roger Keil. 2006. *The Global Cities Reader.* London and New York: Routledge.

Bresson, Alain. 2015. *The Making of the Ancient Greek Economy: Institutions, Markets, and Growth in the City-States.* New York: Princeton University Press.

Brown, David. 2003. *The State and Ethnic Politics in Southeast Asia.* London: Routledge.

Brown, Lance Jay, and David Dixon. 2014. *Urban Design for an Urban Century: Shaping More Livable, Equitable, and Resilient Cities.* Hoboken: Wiley.

Brunn, Stanley D., Jack Williams, and Donald J. Zeigler. 2003. *Cities of the World: World Regional Urban Development.* Lanham: Rowman & Littlefield.

Budd, Eric N. 2004. *Democratization, Development and the Patrimonial State in the Age of Globalization.* Lanham, MD: Lexington Books.

Bulbeck, David. 2004. "Dong-Son Culture in Southeast Asia." In *A Historical Encyclopedia from Angkor Wat to East Timor,* edited by Kiat Gin Ooi, 428–39. Oxford: ABC-CLIO.

Buzsaki, Gyorgy. 2006. *Rhythms of the Brain.* Oxford: Oxford University Press.

Cacciari, Massimo. 2000. "Eupalinos Architecture." In *Architecture Theory Since 1968,* edited by Michael Hays, 394–405. Cambridge, MA: MIT Press.

Cairns, Stephen. 2002. "Troubling Real Estate: Reflection of Urban Form in Southeast Asia." In *Critical Reflections on Cities in Southeast Asia,* edited by Timothy Bunnell, Lisa Barbara Welch-Drummond, and Kong Chong Ho, 101–20. Singapore: Brill & Times Academic Publishing.

Calthorpe, Peter. 1993. *The Next American Metropolis: Ecology, Community, and the American Dream.* New York: Princeton Architectural Press.

———. 2010. *Urbanism in the Age of Climate Change.* Washington, DC: Island Press.

Calthorpe, Peter, and William B. Fulton. 2001. *The Regional City Planning for the End of Sprawl.* Washington, DC: Island.

Carmona, Matthew. 2003. *Public Places, Urban Spaces: The Dimensions of Urban Design.* London: Architectural Press.

Carmona, Matthew. 2014. *Explorations in Urban Design: An Urban Design Research Primer.* Surrey: Ashgate.

Carmona, Matthew, and Steven Tiesdell. 2007. *Urban Design Reader.* London: Architectural Press.

Case, William. 1995. "Malay, Aspects and Audiences of Legitimacy." In *Political Legitimacy in Southeast Asia,* edited by Muthiah Alagappa, 69–107. Stanford: Stanford University Press.

———. 2014. *Routledge Handbook of Southeast Asian Democratization.* London: Routledge.

Castells, Mario. 1996. *The Rise of the Network Society.* Oxford: Oxford Blackwell.

Castell, Mario, and Ida Susser. 2002. *The Castells Reader on Cities and Social Theory.* New York: Wiley-Blackwell.

Catanese, Anthony James, and James C. Snyder. 1979. *Introduction to Urban Planning.* New York: McGraw-Hill.

Cayron, Jun G. 2006. *Stringing the Past: An Archaeological Understanding of Early Southeast Asian Glass Bead Trade.* Diliman Quezon City: University of Philippines Press.

Cervero, Robert. 1998. *The Transit Metropolis: A Global Inquiry.* Washington, DC: Island Press.

Chakwizira, James. 2015. "The Urban and Regional Economy Directing Land Use and Transportation Planning and Development." In *Land Use Management and Transportation Planning,* edited by C. B. Schoeman, 199–225. Hampton and Boston: WIT Press.

Chambert-Loir, Henri. 2005. "Sulalat al-Salatin as Political Myth." *Indonesia* 79: 131–60.

Choras, Daniel D. 2009. *Environmental Science.* Sudbury, MA: Jones & Bartlett Publishers.

Ciancio, Orazio, Piermaria Corona, Francesco Iovino, Giuliano Menguzzato, and Roberto Scotti. 1999. "Forest Management on Natural Basis: Fundamentals and Case Studies in Piermaria Corono." In *Contested Issue of Ecosystem Management,* 121–56. Pennsylvania, PA: Haworth Press.

City of Toronto. 2015. "Toronto Official Plan." Accessed August 16, 2017. https://www1.toronto.ca/wps/portal/contentonly?vgnextoid=03eda07443 f36410VgnVCM10000071d60f89RCRD.

———. 2017a. "Business and Economic Development." Accessed August 17, 2017. https://www1.toronto.ca/wps/portal/contentonly?vgnextoid= 41e067b42d853410VgnVCM10000071d60f89RCRD&vgnextchannel= 57a12cc817453410VgnVCM10000071d60f89RCRD.

———. 2017b. "Get Involved & How Government Works." Accessed August 17, 2017. https://www1.toronto.ca/wps/portal/contentonly?vgnextoid= 8aa4dec8d05bb510VgnVCM10000071d60f89RCRD.

———. 2017c. "Toronto Green Standard Checklist." Accessed August 17, 2017. https://www1.toronto.ca/City%20Of%20Toronto/City%20Planning /Developing%20Toronto/Files/pdf/TGS/2017TGS_LowRise_Checklist. pdf.

———. 2017d. "Toronto Green Standard: Making a Sustainable City Happen." Accessed August 17, 2017. https://www1.toronto.ca/City%20Of% 20Toronto/City%20Planning/Developing%20Toronto/Files/pdf/TGS/2017 TGS_LowRise_Standard.pdf.

City's Manager Office_City of Toronto. 2015. "2015–2018 Strategic Plan." Accessed August 14, 2017. https://www1.toronto.ca/City%20Of% 20Toronto/Equity,%20Diversity%20and%20Human%20Rights/Divisional %20Profile/Policies%20-%20Reports/A1503399_Strat_Plan_web.pdf.

Clearly, Mark, and Chuan Goh Kim. 2000. *Environment and Development in the Straits of Malacca.* London: Routledge.

Cohoone, Lawrence. 2002. *Civil Society: The Conservative Meaning of Liberal Politics.* Malden: Wiley-Blackwell.

Cornelius, Nelarine. 2001. *Human Resource Management: A Managerial Perspective.* London: Thompson.

Coupland, Andy. 1997. *Reclaiming the City: Mixed Use Development.* London: Spon.

Cowan, Alexander, and Jill Steward. 2007. *The Cities and the Senses: Urban Culture Since 1500.* Surrey: Ashgate.

Creswell, Tim. 2014. *Place: A Short Introduction.* Oxford: Blackwell.

Crough, Harold. 1978. *The Army and Politics in Indonesia.* Ithaca: Cornell University Press.

Crouch, Harold. 1996. *Government and Society in Malaysia.* Ithaca, NY and London: Cornell University Press.

Cullingworth, J. Barry. 1987. *Urban and Regional Planning in Canada.* London and New York: Routledge.

Curren, Randall. 2000. *Aristotle on the Necessity of Public Education.* Lanham: Rowman & Littlefield.

D'Angour, Armand. 2011. *The Greeks and the New: Novelty in Ancient Greek Imagination and Experience.* Cambridge: Cambridge University Press.

Darling, Janina. 2004. *Architecture of Ancient Greece: An Account of Its Historic Development.* New York: Biblo & Tannen.

David, Bruno, and Meredith Wilson. 2002. *Inscribed Landscape: Marking and Making Place.* Honolulu, HI: University of Hawaii Press.

Dawson, Ashley, and Brent Hayes Edwards. 2004. *Global Cities of the South.* Durham: Duke University Press.

Delanty, Gerald. 2000. *Citizenship in a Global Age: Society Culture Politics.* Buckingham and Philadelphia: Open University Press.

De Witt, Dennis. 2007. *History of the Dutch in Malaysia.* Kuala Lumpur: Nutmeg Publishing.

Dillon, Sheila. 2006. *Ancient Greek Portrait Sculpture: Contexts, Subjects, and Styles.* Cambridge: Cambridge University Press.

Dinsmoor, William Bell, and William James Anderson. 1973. *The Architecture of Ancient Greece: An Account of Its Historic Development.* Cheshire, CT: Biblo & Tannen Publishers.

Di Piazza, Francesca. 2006. *Malaysia in Picture.* Minneapolis: Visual Geography Series.

Dobbins, Michael. 2011. *Urban Design and People.* Hoboken, NJ: Wiley.

Doolittle, Amity Appel. 2005. *Property and Politics in Sabah, Malaysia: Native Struggle over Land Rights.* Seattle, WA: University of Washington Press.

Dove, Michael R., and Steve Rhee. 2004. "Syahbandar." In *Southeast Asia: A Historical Encyclopedia from Angkor Wat to East Timor,* edited by Keat Gin Ooi, vol. 3, 1286. Singapore: ABC-CLIO.

Doxiadis, Constantinos A. 1964. "The Ancient Greek City and the City of the Present." *Ekistics* 18 (108): 346–64.

Drachenfels, Suzanne Von. 2000. *The Art of the Table: A Complete Guide to Table Setting, Table Manners, and Tableware.* New York: Simon & Schuster.

Drakakish-Smith, D. W. 1987. *The Third World City.* London and New York: Routledge.

Duany, Andres, Elizabeth Plater-Zyberk, and Jeff Speck. 2000. *Suburban Nation: The Rise of Sprawl and the Decline of the American Dream.* New York: The North Point Press, Farrar, Straus and Giroux.

Duany, Andres, Jeff Speck, and Mike Lydon. 2010. *The Smart Growth Manual.* New York, London, and San Francisco: McGraw-Hill.

Elin, Nan. 2013. *Integral Urbanism.* London and New York: Routledge.

El-Khoury, Rodolphe. 2013. *Shaping the City: Studies in History, Theory and Urban Design.* London and New York: Routledge.

El-Khoury, Rodolphe, and Edwards Robbins. 2004/2013. *Shaping the City: Studies in History, Theory, and Urban Design.* London: Routledge.

Ellin, Nan. 2013. *Good Urbanism: Six Steps to Creating Prosperous Places.* Washington, DC: Island Press.

Erhun, Kula. 1998. *History of Environmental Economic Thought.* London: Routledge.

Eriksen, Trond Berg. 1876/1976. *Bios Theoretikos: Notes on Aristotle's Ethica Nicomachea X, 6–8.* Aarhus, DK: Universitetsfod.

Errington, Shelley. 1989. *Meaning and Power in Southeast Asia.* Princeton, NJ: Princeton University Press.

Evangelical Magazine and Missionary Chronicles. 1823. *Memoir of the Late Rev. Milne D.D.* London: Francis Westley.

Evers, Hans-Dieter, and Rudiger Korff. 2000. *Southeast Asian Urbanism: The Meaning and Power of Social Space.* Muenster: LIT.

Fabos, Anita. 2008. *'Brothers' Or Others?: Propriety and Gender for Muslim Arab.* Lanham: Berghahn.

Fader, Steven. 2000. *Density by Design: New Directions in Residential Development.* Washington, DC: Urban Land Institute.

Farr, Douglas. 2008. *Sustainable Urbanism: Urban Design with Nature.* Hoboken, NJ: Wiley.

Fitzgerald, Joseph. 1901. *Word and Phrase: True and False Use in English*. Chicago: A.C. McClurg.

Fitzgerald, Timothy. 2007. *Discourse on Civility and Barbarity: A Critical History of Religion and Related Categories*. Oxford: Oxford University Press.

Florida, Richard. 2008. *Who's Your City?: How the Creative Economy Is Making Where to Live the Most Important Decision of Your Life*. Toronto: Vintage Canada.

Flugel, J. C. 2007. *Man, Moral, and Society*. London: Read Books.

Forth, Gregory. 2009. *A Tale of Two Villages: Hierarchy and Precedence in Keo Dual Organization, Flores Indonesia*. Canberra: ANU Press.

Foucault, Michel. 1985. *History of Sexuality, Volume 2: The Use of Pleasure*. Translated by Robert Hurley. New York: Vintage.

French, Jere Stuart. 1978. *Urban Space: A Brief History of the City Square*. Dubuque: Kendall/Hunt Publishing.

Friedman, Avi. 2006. *Sustainable Residential Development: Planning and Design for Green Neighborhoods*. New York: McGraw-Hill.

Fulton, William B. 1996. *The New Urbanism: Hope or Hype for American Communities?* Cambridge, MA: Lincoln Institute of Land Policy.

Garver, Eugene. 1994. *Aristotle's Rhetoric: An Art of Character*. Chicago: University of Chicago Press.

Gates, Charles. 2003. *Ancient Cities: The Archaeology of Urban Life in the Ancient Near East and Egypt, Greece and Rome*. New York and London: Routledge.

Geddes, Patrick. 1915. *Cities in Evolution*. London: William & Norgate.

Geertz, Clifford. 1980. *Negara: The Theatre State in the Nineteenth Century Bali*. Princeton: Princeton University Press.

———. 1983. *Local Knowledge*. New York: Basic Book.

Gepts, Paul. 2012. *Biodiversity in Agriculture: Domestication, Evolution, and Sustainability*. Cambridge, UK: Cambridge University Press.

Gernet, Louis. 1981. *The Anthropology of Ancient Greece*. Baltimore: Johns Hopkins University Press.

Gidden, Anthony. 1990. *The Consequences of Modernity*. Stanford: Stanford University Press.

———. 1998. *The Third Way: The Renewal of Social Democracy*. Cambridge, UK: Polity Press.

Goh, Robbie G. H., and Brenda S. A. Yeoh. 2003. *Theorizing the Southeast Asian City as Text: Urban Landscapes, Cultural Documents, and Interpretive Experiences*. Singapore: World Scientific Publishing.

Goldberger, R. 2007. "The Rise of Private City." In *Urban Design Reader*, edited by Matthew Carmona and Steve Tiesdell, 170–76. London: Architectural Press.

Goldberg-Miller, Shoshanah B. D. 2017. *Planning for a City of Culture: Creative Urbanism in Toronto and New York.* New York and London: Routledge/Taylor & Francis.

Gomez, Alberto G. 2007. *Modernity and Malaysia: Settling the Menraq Forest Nomads.* London and New York: Routledge.

Gonzales, Francisco. 2006. "Beyond of Beneath Good and Evil? Heidegger's Purification of Aristotle's Ethics." In *Heidegger and the Greeks: Interpretative Essays*, edited by Drew Hyland and Panteleimon Manossakis, 127–56. Bloomington: Indiana University Press.

Goon, Kitt Chin. 1984. *Chinese Geomancy, Feng Shui.* Brisbane: University of New South Wales Press.

Gouldson, Andrew, and Peter Roberts. 2000. *Integrating Environment and Economy: Strategies for Local and Regional Government.* London and New York: Routledge.

Grant, Jill. 2006. *Planning the Good Community: New Urbanism in Theory and Practice.* London and New York: Routledge/Taylor & Francis.

Grime, Charles E. 1996. "Indonesian, an Official Language of Multi Ethnical Nation." In *Atlas of Languages of Intercultural Communication in Pacific, Asia, America*, edited by Stephen Adolphe Wurm, Peter Mulhauser, and Darell T. Tyron. Berlin: Walter de Gruyter.

Grimley, Naomi, and BBC. 2016. "Identity 2016: 'Global Citizenship' Rising, Poll Suggests." Accessed April 28, 2016. http://www.bbc.com/news/world-36139904.

Gudeman, Stephen, and Chris Hann. 2015. *Oikos and Market: Explorations in Self-Sufficiency After Socialism.* New York and Oxford: Berghahn.

Guerrero, Laura K., and Lisa Farenelli. 2009. "The Interval of Verbal and Nonverbal Codes." In *21st Century Communication: A Reference Handbook*, edited by William F. Eadie, vol. 1, 239–48. London: Sage.

Gugler, Josef. 2004. *World Cities Beyond the West: Globalization, Development and Inequality.* Cambridge, UK: Cambridge University Press.

Gupta, Arun Das. 2001. "The Maritime Trade in Indonesia 1500–1800." In *South East Asia Colonial History Before 1800*, edited by Paul H. Kratoska, 92–101. London: Taylor & Francis.

Habermas, Jurgen. 1989. *The Structural Transformation of the Public Sphere: An Inquiry into a Category of Bourgeois Society.* Translated by Thomas Burger and Frederick Lawrence. Cambridge, MA: MIT Press.

Hakim, Besim S. 2014. *Mediterranean Urbanism: Historic Urban/Building Rules and Processes*. Dordrecht: Springer.

Hall, Kenneth R. 2011. *A History of Early Southeast Asia: Maritime Trade and Societal Development 100–1500*. Lanham, MD: Rowman & Littlefield.

Hall, Peter. 1975. *Urban and Regional Planning*. Newton Abbott, London, and Vancouver: David & Charles.

———. 2001/2007. "The City of Theory." In *The City Reader*, edited by Richard T. LeGates and Frederic Stout, 431–444. London: Routledge.

———. 2014. *Cities of Tomorrow: An Intellectual History of Urban Planning and Design Since 1880*. Hoboken, NJ: John Willey & Sons.

Hall, Peter Geoffrey. 1998. *Cities in Civilization*. London: Pantheon Books.

Hall, Thomas. 1997. *Planning Europe's Capital Cities: Aspects of Nineteenth-Century Urban Development*. London: E & F. N. Spon.

Hansen, Mogens Herman. 1983. *The Athenian Ecclesia: A Collection of Articles 1976–1983*. Copenhagen: Museum Tusculanum Press.

———. 1991. *The Athenian Democracy in the Age of Demosthenes: Structure, Principles, and Ideology*. Translated by J. A. Crook. Norman: Oklahoma University Press.

———. 1993. *The Ancient Greek City-State: Symposium on the Occasion of the 250th Anniversary of the Royal Danish Academy of Sciences and Letters July, 1–4 1992*. Copenhagen: Kongelige Danske Videnskabenes Selskab.

———. 1997. *The Polis as an Urban Centre and as a Political Community*. Copenhagen: Royal Danish Academy.

———. 2002. *Even More Studies in Ancient Greek Polis*. Edited by Thomas Heine Nielsen. Stuttgart: Franz Steiner.

Hansen, Mogens Herman, and Tobias Fischer Hansen. 1994. "Monumental Political Architecture in Archaic and Classical Greek Poleis." In *From Political Architecture to Stephanus Byzantius*, edited by David Whitehead, 23–89. Stuttgart: Franz Steiner.

Harborfront Center. 2015. "Harborfront Center." Accessed August 17, 2017. http://www.harbourfrontcentre.com/.

Harris, Edward M., David M. Lewis, and Mark Woolner. 2015. *The Ancient Greek Economy: Markets, Households and City-States*. Cambridge, UK: Cambridge University Press.

Harris, Nathaniel. 2004. *History of Ancient Greece*. New York: Barnes & Noble.

Harvey, David. 1998. *The Urban Experience*. Oxford: Basil Blackwell.

———. 2008. "The Right to the City." January. Accessed March 1, 2016. http://davidharvey.org/media/righttothecity.pdf.

Haverfield, Francis-John. 2009. *Ancient Town Planning*. Auckland, New Zealand: BiblioLife LLC.

Haworth, Alan. 2004. *Understanding the Political Philosophers: From Ancient to Modern Times*. London: Routledge.

Heidegger, Martin. 1927/2010. *Being and Time*. Edited by Dennis J. Schmidt. Translated by Joan Stambaugh. Albany, NY: SUNY Press.

———. 1977. *Basic Writings*. Edited by David Farrell Krell. Translated by Adolf Hofstadter. San Francisco: Harper & Row.

———. 1991. *Nietzsche Volume I & II*. Edited by David Farrell Krell. Translated by David Farrell Krell. San Francisco: HarperSanFrancisco.

———. 2007. *Die Kunst und der Raum*. Frankfurt am Main: Vittorio Klostermann.

———. 2012. *Bremen and Freiburg Lectures: Insight into That Which Is and Basic Principles of Thinking*. Translated by Andrew J. Mitchell. Bloomington and Indianapolis, IN: Indiana University Press.

Heller, Steven, and Marie Finnamore. 1997. *Design Culture: An Anthology of Writing from the AIGA Journal of Graphic Design*. New York: Allworth Press.

Hester, Randolph T. 2006. *Design for Ecological Democracy*. Cambridge, MA: MIT Press.

Hestflatt, Kristin. 2003. "The House as a Symbol of Identity Among the Baba Melaka." In *The House in Southeast Asia: A Changing Social, Economic, and Political Domain*, edited by Stephen Sparkes and Signe Howell, 67–82. London: Routledge.

Hill, Hal, Siew Yean Tham, and Ragayah Haji Mat Zin. 2013. *Malaysia's Development Challenges: Graduating from the Middle*. London and New York: Routledge.

Hilley, John. 2001. *Malaysia: Mahathirism, Hegemony and the New Opposition*. London and New York: Zed Books.

Hinde, Robert A. 1972. *Non-verbal Communication*. Cambridge: Cambridge University Press.

Hodge, Gerad, and David L. A. Gordon. 2008. *Planning Canadian Communities: An Introduction to Principles, Practice, and Participants*. 5th ed. Toronto: Thomson Nelson.

Hofstede, Geert. 2001. *Culture's Consequences: Comparing Values, Behaviors Institutions, and Organization Across Nations*. London: Sage.

Holston, James. 1989. *The Modernist City: An Anthropological Criticism of Brasilia*. Chicago: University of Chicago Press.

Hooper, Finley. 1978. *Greek Realities, Life and Thought in Ancient Greece*. Detroit: Wayne State University Press.

Huntington, Samuel P. 1993. "The Clash of Civilization?" *Foreign Affairs* 72 (3): 21–49.

Husserl, Edmund. 1962/2002/2014. *Ideas, General Introduction to Pure Phenomenology*. Translated by W. Boyer Gibson. London and New York: Routledge.

———. 1901–1902/2001. *Logical Investigations*, vol. 1. Translated by N. Findlay. London: Routledge.

———. 1970. *The Crises of European Sciences and Transcendental Phenomenology*. Translated by David Carr. Evanston: Northwestern University.

Hussin, Nordin. 2007. *Trade and Society in the Straits of Melaka: Dutch Melaka and English*. Copenhagen: NIAS.

Ikerd, John E. 2005. *Sustainable Capitalism: A Matter of Common Sense*. Sterling, VA: Kumarian Press.

Iskandar, Yusoff. 1992. *The Malay Sultanate of Malacca: A Study of Various Aspects of Malacca in the 15th and 16th Centuries in Malaysian History*. Kuala Lumpur: Dewan Bahasa & Pustaka.

Ismail, Rahil, Brian J. Shaw, and Giok Ling Ooi. 2009. *Southeast Asian Culture and Heritage in a Globalizing World*. Surrey: Ashgate.

Jacobs, Jane. 1961. *The Death and Life of the Great American Cities*. New York: Random House.

Jaeger, Werner Wilhelm. 1939/1986. *Paideia: The Ideals of Greek Culture*. Translated by Gilbert Highet. Oxford: Oxford University Press.

James, Paul, and Manfred B. Steger. 2016. "Globalization and Global Consciousness: Levels of Connectivity." In *Global Culture: Consciousness and Connectivity*, edited by Roland Robertson and Didem Buhari-Gulmez, 21–40. Abingdon and New York: Ashgate/Routledge.

Jarq-Hergoualc'h, Michel. 2011. *The Malay Peninsula: Crossroad of the Maritime Silk Road 100 BC–1300 AD*. Leiden: Brill.

Jencks, Chris. 2004. *Urban Culture: Critical Concepts in Literary and Cultural Studies*. London: Routledge/Taylor & Francis.

Jerke, Dennis, Douglas R. Porter, and Terry J. Lassar. 2008. *Urban Design and the Bottom Line: Optimizing the Return on Perception*. Washington, DC: Urban Land Institute.

Joeman, Boyd Dyonisius. 2013. *Iskandar Malaysia Smart City Framework*. March 21–22. http://hls-esc.org/documents/4hlsesc/2C%20-%20Iskandar.pdf.

Johnston, Sarah Iles. 2004. *Religions of the Ancient World: A Guide*. Cambridge, MA: Harvard University Press.

Jomo, Kwame Sundaram, and Sue Ngan Wong. 2012. *Law Institutions and Malaysian Economic Development*. Singapore: NUS Press.

Jordan, Jennifer. 2003. "Collective Memory and Locality in Global Cities." In *Global Cities, Cinema, Architecture, and Urbanism in a Digital Age*, edited by Linda Krause and Patricia Petro, 31–48. Fredericksburg: Rutgers University Press.

Jordan, Michael. 1993. *Ancient Concepts of Philosophy*. London: Routledge.

Julier, Guy. 2000. *The Culture of Design, Culture, Media and Identities Series*. London: Sage.

Kagan, Donald. 1991. *Pericles of Athens and the Birth of Democracy*. New York: Free Press.

Kalimitziz, Kostas. 2000. *Aristotle on Political Enmity and Disease: An Inquiry into Stasis*. Albany, NY: SUNY Press.

Kant, Immanuel. 1781/1990. *Critique of Pure Reason*. Translated by Mortimer J. Adler. London: Encyclopedia Britannica.

Katz, Peter, Vincent Scully, and Todd W. Bressi. 1994. *The New Urbanism: Toward an Architecture of Community*. New York: McGraw-Hill.

Kelbough, Douglas, and Kit Krankel McCullough. 2008. *Writing Urbanism: An ACSA Publication*. London and New York: Routledge.

Kennedy, Roger. 2014. *The Psychic Home: Psychoanalysis, Consciousness and the Human Soul*. London and New York: Routledge.

King, Anthony D. 2015. *Global Cities*. Routledge Library Editions: Economic Geography. London and New York: Routledge.

King, Ross. 2008. *Kuala Lumpur and Putrajaya: Negotiating Urban Space in Malaysia*. Singapore: NUS Press.

Kipfer, Barbara Ann. 2000. *Encyclopedic Dictionary of Archaeology*. New York: Springer.

Klimasmith, Betsy. 2005. *At Home in the City: Urban Domesticity in American Literature and Culture 1850–1930*. Lebanon: University of New Hampshire Press.

Knauf, Bruce. 1996. *Genealogies for the Present in Cultural Anthropology*. London: Routledge.

Knaur, Hadev, and Mizan Hitam. 2010. "Sustainable Living: An Overview from Malaysian Perspective." In *Towards a Liveable and Sustainable Urban Environment, Eco Cities in East Asia*, edited by Lye Liang Fook and Chen Gang, 159–78. Singapore: World Scientific.

Kolb, Robert W. 2008. *Encyclopedia of Business Ethics and Society*, vol. 1. London: Sage.

Kolson, Kenneth. 2003. *Big Plans: The Allure and Folly of Urban Design*. Baltimore: Johns Hopkins University Press.

Konstan, David. 1997. *Friendship in the Classical World.* Cambridge: Cambridge University Press.

Kostof, Spiro. 1991. *The City Shaped: Urban Patterns and Meanings Through History.* London: Thames & Hudson.

Kratoska, Paul. 2001. *South East Asia, Colonial History: Imperialism Before 1800.* London: Taylor & Francis.

Krause, Linda, and Petrice Petro. 2003. *Cinema, Architecture, and Urbanism in Digital Age.* New Brunswick, NJ and London: Rutdgers University Press.

Kraut, Richard, and Steven Skultety. 2005. *Aristotle's Politics: Critical Essays.* Lanham, New York, and Toronto: Rowman & Littlefield.

Krell, David Farell. 1988. "Knowledge Is Remembrance; Diotima's Instruction." In *Post-structuralist Classics,* edited by Andrew E. Benjamin, 160–72. London: Routledge.

Krier, Leon. 1984. "Houses, Palaces, Cities." Edited by Demetri Porphyrios. Architectural Design 7/8.

Kulke, Hermann. 1991. "Epigraphical Reference of City and State in Early Indonesia." *Indonesia* 52: 3–22.

Lang, Jon. 1994. *Urban Design: The American Experience.* Hoboken, NJ: Wiley.

———. 2005. *Urban Design: A Typology of Procedures and Products.* London: Architectural Press.

Lange, Matthew. 2006. *Lineage of Despotism and Development: British Colonialism and State Power.* Chicago: University of Chicago Press.

Larice, Michael, and Elizabeth Macdonald. 2013. *The Urban Design Reader.* London and New York: Routledge.

Lechner, Frank J., and John Boli. 2012. *The Globalization Reader.* Malden, MA: Wiley-Blackwell.

Lee, Boon Thong. 1999. "Emerging Urban Trend and Globalizing Economy in Malaysia." In *Emerging Cities in Pacific Area,* edited by Fu-Chen Lo and Yeong, 336–48. Tokyo: UNU Press.

LeGates, Richard T., and Frederic Stout. 2003. *The City Reader: Routledge Urban Reader Series.* London: Routledge.

Leifer, Michael. 1978a. *Malacca.* Dordrecht: Brill.

———. 1978b. *Malacca, Singapore, and Indonesia.* Dordrecht: Brill.

Lemon, James T. 2008. *Liberal Dreams and Nature's Limits: Great Cities of North America Since 1600.* Eugene, OR: Wipf & Stock.

Lems, Kristin, Leach D. Miller, and Tenena M. Soro. 2009. *Teaching Reading to English Language Learners: Insights from Linguistics.* New York: Guilford Press.

Lefebvre, Henri. 1991. *The Production of Space.* Translated by Donald Nicholson-Smith. Malden: Blackwell.

Levine, Allan. 2014. *Toronto: Biography of a City*. Madeira Park, BC: Douglas & McIntyre.

Levi-Strauss, Claude. 1961. *Tristes Tropiques*. New York: Atheneum.

———. 1963. *Structural Anthropology*. Translated by Claire Jacobson and Cladia Schoepf. New York: Basic Books.

———. 1966. *The Savage Mind*. Translated by John Weightman and Doreen Weightman. Chicago: University of Chicago Press.

Levy, John M. 2009. *Contemporary Urban Planning*. Upper Saddle River, NJ: Pearson Prentice Hall.

Liddel, Peter. 2007. *Civic Obligation and Individual Liberty in Ancient Athens*. Oxford: Oxford University Press.

Lim, Jee Yuan. 2011. *The Malay House: Principles to Building Simple and Beautiful Homes for Comfort and Community*. East Peterburg: Fox Chapel.

Locke, John. 1841/2007. *Essay Concerning Human Understanding*. Oxford: Oxford University Press.

Loukaitou-Sideris, Anastasia, and Tridib Banerjee. 1998. *Urban Design Downtown: Poetics and Politics of Form*. Berkeley: University of California Press.

Lovett, Gary, Clive G. Jones, Monica G. Turner, and Kathleen C. Weathers. 2005. *Ecosystem Function in Heterogeneous Landscapes*. Berlin: Birkhauser.

Lynch, Kevin. 1981. *A Theory of Good City Form*. Cambridge: MIT Press.

MacArthur, Robert Halmer, and Edward O. Wilson. 1967. *The Theory of Island Biogeography*. Princeton, NJ: Princeton University Press.

Mackey, Eva. 1999. *House of Difference: Cultural Politics and National Identity in Canada*. London: Routledge.

Mackintosh, Phillip Gordon. 2017. *Newspaper City: Toronto's Street Surfaces and the Liberal Press, 1860–1935*. Toronto: University of Toronto Press.

Madanipour, Ali. 2014. *Urban Design, Space and Society*. London and New York: Palgrave Macmillan.

Maguin, Pierre-Yves. 2000. "City-States and City-State Cultures in Pre 15th Century Southeast Asia." In *A Comparative Study of City-State Cultures: An Investigation*, edited by Mogens Hermann Hansen, 409–30. Copenhagen: University of Copenhagen.

Mandel, Sumit K. 2001. "Boundaries and Beyond, Whither the Cultural Bases of Political Community in Malaysia." In *The Politics of Multiculturalism: Pluralism and Citizenship in Malaysia Singapore Indonesia*, edited by Robert W. Hefner, 141–64. Honolulu, HI: University of Hawaii Press.

Manguin, Piere-Yves. 2000. "City States and City State Culture in the Pre 15 Century Southeast Asia." In *A Comparative Study of Thirty City-State Cul-*

tures: An Investigation, edited by Mogens Herman Hansen, vol. 21, 409–30. Copenhagen: Kongelige Danske Videnskabernes Selskab.

Manzi, Tony, Karen Lucas, Tony Lloyd Jones, and Judith Allen. 2010. *Social Sustainability in Urban Areas: Communities, Connectivity and the Urban Fabric.* London: Routledge.

Martin-McAuliffe, Samantha L., and Daniel M. Millette. 2017. *Ancient Urban Planning in the Mediterranean: New Research Directions.* New York and London: Routledge.

Massey, Doreen. 1994. *Space, Place, and Gender.* Minneapolis: University of Minnesota Press.

Massey, Gerald, and Charles S. Finch. 1998. *The Natural Genesis.* Baltimore: lack Classic Pers.

McCarthy, Justin. 2014. "Gallup: Trust in Mass Media Returns to All-Time Low." September 17. Accessed March 1, 2016. http://www.gallup.com/poll/176042/trust-mass-media-returns-time-low.aspx.

McClain, James L., and Osamu Wakita. 1999. *Osaka: The Merchants' Capital of Early Modern.* Ithaca, NY: Cornell University Press.

Mcclelland, J. S. 2015. *A History of Political Thought.* London and New York: Routledge.

McGee, T. G. 1967. *The Southeast Asian City: A Social Geography of the Primate Cities of Southeast Asia.* New York: Praeger.

McGee, T. G. 2002. "Reconstructing 'The Southeast Asian City in an Era of Volatile Globalization'." In *Critical Reflections on Cities in Southeast Asia,* edited by Tim Bunnell, Lisa B. W. Drummond, and K. C. Ho, 31–53. Singapore: Brill & Times.

McGee, T. G., and Ira M. Robinson. 2011. *Mega Urban Regions of Southeast Asia.* Vancouver: UBC Press.

McIntosh, Jane, and Clint Twist. 2001. *Civilizations: Ten Thousand Years of Ancient History.* London: BBC Worldwide.

McLennan, Jason F. 2004. *The Philosophy of Sustainable Design.* Bainbridge Island, WA: Ecotone.

Meer, N. C. Van Setten van der. 1979. Sawah Cultivation in Ancient Java, Aspects of Development during the Indo-Javanese Period 5th century. Oriental Monograph Series No. 22. Canberra: ANU Faculty of Asian Studies.

Mega, Voula. 2005. *Sustainable Development, Energy and the City: A Civilization of Visions and Actions.* New York and Berlin: Springer.

Miksic, John N. 2004. "The Classical Cultures of Indonesia." In *Southeast Asia from Prehistory to History,* edited by Ian Glover and Peter Belwood, 234–56. London and New York: Routledge.

Miles, Malcolm, Tim Hall, and Iain Borden. 2004. *The City Cultures Reader.* New York: Routledge.

Miraftab, Faranak, and Neema Kudva. 2015. *Cities of the Global South Reader.* London and New York: Routledge.

Mitchell, Don. 2003. *The Right to the City, Social Justice and the Fight for Public Space.* New York: Guilford.

Moller, Astrid. 2000. *Trade in Archaic Greece.* Oxford: Oxford University Press.

Moran, Dermot. 2000. *Introduction to Phenomenology.* London: Routledge.

Morris, A. E. J. 1972/2013. *History of Urban Form Before the Industrial Revolution.* London and New York: Routledge.

Morris, Ian, and Barry B. Powell. 2006. *The Greeks, History, Culture and Society.* New York: Pearson Prentice Hall.

Morris, Ian, and Barry B. Powell. 2010. *The Greeks: History, Culture, and Society.* 2nd ed. Englewood Cliffs: Prentice Hall.

Morrison, Kathleen. 2002. "Pepper in the Uphills: Upland and Lowland Exchange and the Intensification of the Spice Trade." In *Forager-Traders in South and Southeast Asia, Longterm Histories,* edited by Kathleen D. Morrison and Laura L. Junker, 105–30. Cambridge: Cambridge University Press.

Moughtin, Cliff, Rafael Cuesta, Christine Sarris, and Paola Signoretta. 1999/2003. *Urban Design: Method and Techniques.* Oxford and London: Architectural Press.

Mumford, Lewis. 1961. *The City in History: Its Origins, Its Transformations and Its Prospects.* New York: Harcourt, Brace and World.

Munch, Richard, and Neil J. Smeiser. 1992. *Theory of Culture.* Berkeley: University of California Press.

Municipality of Taiping, Perak. 2015. "Taiping Background." May 19. Accessed May 21, 2016. http://www.mptaiping.gov.my/en/latar-belakang-taiping.

Murray, Oswyn. 1991. *The Greek City: From Homer to Alexander.* Oxford: Clarendon Press.

Mustafa, Maizatun. 2011. *Environmental Law in Malaysia.* AH Alphen aan den Rijn: Wolters Kluwer.

Nagata, Judith A. 1975. *Contributions to Asian Studies: Pluralism in Malaysia: Myth and Reality: A Symposium on Singapore and Malaysia.* Leiden: Brill.

Nagel, Mechthild. 2002. *Masking the Abject: A Genealogy of Play.* Lanham: Lexington.

Nagle, D. Brendan. 2006. *The Household as the Foundation of Aristotle's Polis.* Cambridge: Cambridge University Press.

Nas, Peter J. M., and Reynt J. Sluis. 2002. "In Search of Meaning." In *The Indonesian Town Revisited,* edited by Peter J. M. Nas, 130–46. Muenster: LIT.

Neuliep, James William. 2009. *Intercultural Communication: A Contextual Approach*. London: Sage.

Newbold, Thomas John. 1839. *Political and Statistical Account of the British Settlements in the Straights of Malacca*. London: John Murray.

Newitt, M. D. D. 1986. *The First Portuguese Colonial Empire*. Exeter: University of Exeter Press.

Newman, Peter. 1996. "Reducing Automobile Dependence." *Environment and Urbanization* 8 (1): 67–72.

Nielsen, Thomas Heine. 2002. *Even More Studies in Ancient Greek Polis*. Stuttgart: Frans Steiner Verlag.

Nolon, John R. 2006. *Compendium of Land Laws for Sustainable Development*, 97–111. Cambridge: Cambridge University Press.

North, Douglas. 1991. "Institutions." *Journal of Economic Perspectives* 5 (1): 97–112.

Ober, Josiah. 2011. *Political Dissent in Democratic Athens: Intellectual Critics of Popular Rule*. New York: Princeton University Press.

Olson, Kevin. 2008. "Governmental Rationality and Popular Sovereignty." In *No Social Science Without Critical Theory*, edited by Harry F. Dahms, vol. 25, 329–52. Bingley: Emerald.

Ooi, Keat Gin. 2004. *Southeast Asia: A Historical Encyclopedia, from Angkor Wat to East Timor*. St. Barbara, CA: ABC-CLIO.

———. 2009. *Historical Dictionary of Malaysia*. Lanham: Scarecrow.

Orr, David. 2004. *The Nature of Design: Ecology, Culture, and Human Intention*. Oxford and London: Oxford University Press.

Orrieux, Claude, and Pauline Schmitt. 1999. *A History of Ancient Greece*. London: Blackwell.

Osman, Mohammad Taib. 1985. *Malaysian World View*. Singapore: Institute of Southeast Asian Studies.

Ostler, Thomas. 2014. *Downtown Toronto: Trends Issues Intensification*. Toronto, Ontario: City of Toronto, City Planning - Toronto and East York District.

Owen, Norman G. 1999. "Economic Social Change." In *Cambridge History of Southeast Asia from the World War II to the Present*, edited by Nicholas Tarling, 139–98. Cambridge, UK: Cambridge University Press.

Pagden, Anthony. 2002. *The Idea of Europe: From Antiquity to European Union*. Cambridge: Cambridge University Press.

Pagliaro, Jennifer. 2017. "Onward and Upward." February 17. http://projects. thestar.com/ontario-municipal-board-reform/onward-upward/.

Palmer, David, and Michael Joll. 2011. *Tin Mining in Malaysia 1800–2000: The Osborne & Chappel Story*. Kuala Lumpur: Perpustakaan Negara.

Parfect, Michael, and Gordon Power. 1997. *Planning for Urban Quality: Urban Design in Towns and Cities.* London: Routledge.

Parnell, Susan, and Sophie Oldfield. 2014. *The Routledge Handbook on Cities of the Global South.* London and New York: Routledge.

Passell, Aaron. 2013. *Building the New Urbanism: Places, Professions, and Profits in the American Metropolitan Landscape.* London and New York: Routledge.

Pateman, John. 2013. *Fort Pitt to Fort William.* Thunder Bay: Pateran Press.

Partridge, E. Staff, and Eric Patridge. 1977. *Origin of Etymology Dictionary of Modern English.* 4th ed. London: Routledge.

Partridge, Lesley. 1999. "Creating Competitive Advantage with HRM." www.selectknowledge.com: Select Knowledge.

Pieris, Amona. 2009. *Hidden Hands and Divided Landscapes: A Penal History of Singapore's Plural Society.* Honolulu, HI: University of Hawaii Press.

Pires, Tome, Armando Cortesao, and Rodrigues Francisco. 1990. *The Summa Oriental of Tome Pires.* Kuala Lumpur: Asian Educational Service.

Pomeroy, Sarah B., Stanley M. Burnstein, and Walter Donlan. 1998. *Ancient Greece: A Political, Social, and Cultural History.* New York: Oxford University Press.

Preston, Laura, and Sara Owen. 2009. *Inside the City in the Greek World.* Oxford: Oxbow.

Price, Jonathan J. 2001. *Thucydides and Internal War.* Cambridge: Cambridge University Press.

Pride, William M. 1992. *Business.* Boston: Houghton Mifflin.

Puffert, Douglas J. 2009. *Tracks Across Continents, Paths Through History: The Economic Dynamics of Standardization in Railway Gauge.* Chicago: University of Chicago Press.

Punter, J. V. 2007. "Urban Design as Public Policy: Best Practice Principles for Design Review and Development Management." *Journal of Urban Design* 12 (2): 167–202.

Punter, John. 2009. *Urban Design and the British Urban Renaissance.* London and New York: Routledge.

Pye, Lucian W., and Mary W. Pye. 2009. *Asian Power and Politics: The Cultural Dimensions of Authority.* Cambridge, MA: Harvard University Press.

Rae, Douglas W. 2005. *Urbanism and Its End.* New Haven: Yale University Press.

Ramirez-Lovering, Diego. 2008. *Opportunistic Urbanism.* Melbourne: MIT Press.

Rappa, Antonio, and Lionel Wee. 2006. *Language Policy and Modernity in Southeast Asia: Malaysia, the Philippines.* Singapore: Springer.

Ratcliffe, John, and Michael Stubbs. 2004. Urban Planning and Real Estate Development, Volume 8 of the Natural and Built Environment Series. London: Taylor & Francis.

Ray, Christopher. 1991. *Space, Time, and Philosophy.* New York: Routledge.

Ray, Huang. 1982. *1587, a year of No Significance: The Ming Dynasty in Decline.* New Haven: Yale University Press.

Redway, Alan. 2014. *Governing Toronto: Bringing Back the City That Worked.* Victoria: Friesen.

Reid, Anthony. 2000. *Southeast Asia in the Early Modern Era: Trade, Power, and Belief.* Ithaca, NY: Cornell University Press.

———. 2002. *Southeast Asia in the Age of Commerce, 1450–1680: The Lands Below the Winds.* New Haven: Yale University Press.

Ricklefs, Merle Calvin. 2001/2008. *A History of Modern Indonesia Since c. 1200.* Stanford: Stanford University Press.

Riddle, Robert. 2004. *Sustainable Urban Planning.* London: Routledge.

Riffkin, Rebecca. 2015. "Gallup: Americans' Trust in Media Remains at Historical Low." September 28. Accessed March 1, 2016. http://www.gallup.com/poll/185927/americans-trust-media-remains-historical-low.aspx.

Rimmer, Peter James, and Howard W. Dick. 2009. *The City in Southeast Asia: Patterns, Processes and Policy.* Singapore: NUS Press.

Roberts, Lance W., Rodney A. Clifton, and Barry Ferguson. 2005. *Recent Social Trends in Canada, 1960–2000, Comparative Charting of Social Change.* Montreal: McGill-Queen University Press.

Robertson, Roland, and Kathleen E. White. 2003. *Globalization: Critical Concepts in Sociology.* London and New York: Routledge.

Robinson, Charles Alexander. 1959. *Athens in the Age of Pericles.* Norman: University of Oklahoma Press.

Roth, Leland M., and Amanda C. Roth Clark. 2014. *Understanding Architecture: Its Elements, History, and Meaning.* Boulder: Westview.

Roy, Ananya, and Jennifer Robinson. 2015. "Global Urbanism and the Nature of Theory." *International Journal* (Willey) 40: 1–11.

Roy, Ananya, and Nezar AlSayad. 2004. *Urban Informality: Transnational Perspectives from Middle East, Latin America and South Asia.* Lanham: Lexington Books.

Rudner, Martin. 1994. *Malaysian Development: A Retrospect.* Montreal: McGill-Queen University Press.

Sack, Robert David. 1986. *Territoriality and Its History; Human Territoriality.* Cambridge: Cambridge University Press.

Sagoff, Mark. 1995. "The Value of Integrity." In *Perspectives on Ecological Integrity*, edited by Laura Westra and John Lemons, 162–76. Dordrecht: Springer Science & Business Media.

Sahlins, Marchall D. 1976. *Culture and Practical Reason*. Chicago: University of Chicago Press.

Saleh, Ismail Mohammad, and Saha Devan Meyanathan. 1993. *The Lesson of East Asia, Malaysia: Growth, Equity, and Structural Transformation*. Washington, DC: The World Bank.

Sanders, William T., and Barbara J. Price. 1968. *Mesoamerica: The Evolution of a Civilization*. New York: Random House.

Sandhu, Kernial Singh, and Paul Wheatley. 1983. *Melaka: The Transformation of a Malay Capital c. 1400–1980*. Oxford: Oxford University Press.

Sarton, George. 1952/1993. *Ancient Science Through the Golden Age of the Greece*. Mineola: Courier.

Sassen, Saskia. 1991/2013. *The Global City: New York, London, Tokyo*. New York: Princeton University Press.

Saunders, Peter. 1995. *Social Theory and the Urban Question*. New York and London: Routledge.

Saw, Swee-Hock, and K. Kesavapany. 2006. *Malaysia, Recent Trends and Challenges*. Singapore: Institute of Southeast Asian Studies.

Schmitt-Pantel, Pauline. 1991. "Collective Activities and the Political in the Greek City." In *The Greek City: From Homer to Alexander*, edited by Oswyn Murray, 199–213. Oxford: Clarendon Press.

Schmitz, Oswald J. 2007. *Ecology and Ecosystem Conservation*. Washington, DC: Island Press.

Schulze, Ernst-Detlef, and Harold A. Mooney. 2012. *Biodiversity and Ecosystem Function*. Berlin, Heidelberg, and New York: Springer Science & Business Media.

Scott, Alen J., and Michael Storper. 2015. "The Nature of the Cities: The Scope and Limit of Urban Theory." *International Journal of Urban and Regional Research* 39 (1): 1–15.

Scott, Bruce R. 2011. *Capitalism: Its Origins and Evolution as a System of Governance*. New York, Dordrecht, and Heidelberg: Springer Business Media.

Scott-Ross, Marcus. 1971. *A Short History of Malacca*. Malacca: Chopmen Enterprise.

Searle, John R. 1983. *Intentionality: An Essay in the Philosophy of Mind*. Cambridge: Cambridge University Press.

Sennet, Richard. 1977. *The Fall of Public Man*. New York: Alfred A. Knopf.

Sewell, John. 1993. *The Shape of the City: Toronto Struggles with Modern Planning*. Toronto, Buffalo, and London: University of Toronto Press.

Shachak, Moshe, James R. Gosz, Stewart T. A. Pickett, and Avi Perevolotsky. 2005. *Biodiversity in Drylands*. New York: Oxford University Press.

Shariff, Abbas Mohammad. 2004. *Adab Orang Melayu*. Kuala Lumpur: Alfa Media.

Sheppard, Eric, Helga Leitner, and Anant Maringanti. 2013. "Provincializing Global Urbanism: A Manifesto." *Journal of Urban Geography* 34: 1–9.

Sheriff, Abdul. 2010. *Dhow Cultures and Indian Ocean: Cosmopolitan, Commerce, and Islam*. New York: Columbia University Press.

Short, John Rennie. 2014. *Urban Theory: A Critical Assessment*. London: Palgrave.

Simanjuntak, Truman. 2006. *Archaeology: Indonesian Perspective: R.P. Soejono Festschrift*. Jakarta: LIPI.

Simmel, Georg. 1950. "The Metropolis and Mental Life." In *The Sociology of Georg Simmel*, edited by D. Weinstein, translated by Kurt Wolff, 409–24. New York: Free Press.

Skinner, Stephen. 1980. *Terrestrial Astrology: Divination by Geomancy*. London: Routledge.

Slomkowski, Paul. 1997. *Aristotle's Topic: What Is a Topos?* Leiden: Brill.

Smith, Peter. 2007. *The Dynamics of Urbanism*. London: Routledge.

Smith, Vincent, and Barbara Lyons Stewart. 2006. *Feng Shui: A Practical Guide for Architects and Designers*. London: Kaplan Publishing.

Smithson, Peter, and Kenneth Atkinson. 2002. *Fundamentals of Physical Environment*. London: Routledge.

Sorkin, Michael. 2009. *The End(s) of Urban Design in Alex Krieger & William S. Saunders, Urban Design*. Minneapolis: University of Minnesota Press.

Stanwick, Sean, Jennifer Flores, and Tom Arban. 2007. *Design City Toronto*. Toronto: Wiley.

Starobinski, Jean. 1993. *Blessings in Disguise, or, the Morality of Evil*. Cambridge, MA: Harvard University Press.

Statistics Canada. 2009-11-20. "Immigration in Canada: A Portrait of the Foreign-Born Population, 2006 Census: Portraits of Major Metropolitan Centres." Accessed August 9, 2017. http://www12.statcan.ca/census-recensement/2006/as-sa/97-557/p24-eng.cfm.

Statistics Malaysia. 2010. "Taburan Penduduk dan Ciri-ciri Azas Demografi." May 2. http://www.statistics.gov.my/portal/index.php?lang=en.

Steele, James. 1997. *Sustainable Architecture: Principles, Paradigms, and Case Studies*. New York: McGraw-Hill.

Steinberg, Philip E., and Rob Shield. 2008. *What Is a City? Rethinking the Urban After Hurricane Katrina.* Atlanta: University of Georgia Press.

Stevenson, Deborah. 2003. *Cities and Urban Cultures.* Philadelphia: McGraw-Hill.

Strootman, Rolf. 2018. "Feasting and Polis Institutions." In *Leiden,* edited by Floris van den Eijnde, Josine Blok, and Rolf Strootman, 273–86. Leiden and Boston: Brill.

Suhaimi, N. H. 1993. "Pre-modern Cities in the Malay Penninsula and Sumatra." *Jurnal Arkeologi Malaysia* 6: 66–77.

Sukuran, Mohammad, and Chin Siong Ho. 2008. "Planning System in Malaysia." Skudai Johor Bahru: Toyohashi University of Technology-UTM Seminar of Sustainable Development and Governance.

Tadjuddin, Mohammad Rasdi, Komarudin Mohammad Ali, Syed Iskandar Syed Arifin, Ra'alah Mohammad, and Gurupiah Mursib. 2005. *The Architectural Heritage of the Malaysian World: Traditional Houses.* Skudai: UTM Press.

Tak, Ming Ho. 2009. *Ipoh.* Ipoh: Perak Academy.

Talen, Emily. 2012. *City Rules: How Regulations Affect Urban Form.* Washington, DC: Island Press.

———. 2013. *Charter of the New Urbanism.* New York: McGraw-Hill.

Tarling, Nicholas. 1999. *The Cambridge History of Southeast Asia.* Cambridge: Cambridge University Press.

Tay, Eddie. 2010. *Colony, Nation, and Globalization: Not at Home in Singaporean and Malaysian Literature.* Hongkong: Hongkong University Press.

Taylor, Claire. 2015. "Social Networks and Social Mobility in Fourth-Century Athens." In *Communities and Networks in the Ancient Greek World,* edited by Claire Taylor and Kostas Vlassopoulos, 35–53. Oxford: Oxford University Press.

Taylor, Claire, and Kostas Vlassopoulos. 2015. *Communities and Networks in the Ancient Greek World.* Oxford: Oxford University Press.

Tey, Tsun Hang. 2011. "Iskandar Malaysia and Malaysia's Dualistic Political Economy." In *Special Zones in Asia Market Economies,* edited by Connie Carter and Andrew Harding. London: Routledge.

The City Of Toronto, Ontario Canada. 2002. "Toronto Official Plan." Accessed August 10, 2017. https://www1.toronto.ca/planning/chapters1-5.pdf#page=17.

The World Bank. 2012. "Indocator." January 19. http://data.worldbank.org/indicator/SP.URB.TOTL.IN.ZS.

Thomas, Carol G. 1988. *Path to Ancient Greece.* Leiden: Brill.

Thomas, Carol. 1993. *Understanding Architecture: Its Elements, History, and Meaning*. Oxford: Westview Press.

Thomas, Luis Filipe Ferreira Reis. 1993. "The Malay Sultanate of Melaka." In *Southeast Asia in Early Modern Era, Trade, Power, and Belief,* edited by Anthony Reid, 69–122. Ithaca, NY: Cornell University Press.

Thomas, Randall. 2003. *Sustainable Urban Design: An Environmental Approach*. London: Taylor & Francis.

Thomas, Ren. 2016. *Planning Canada: A Case Study Approach*. Oxford: Oxford University Press.

Thompson, Eric. 2007. *Unsettling Absences, Urbanism in Rural Malaysia*. Singapore: NUS Press.

Thomsen, Rudi. 1964. *Eisphora: A Study of Direct Taxation in Ancient Athens*. Oslo: Gyldendal.

Tiesdel, Steve, and David Adams. 2011. "Real Estate Development, Urban Design and the Tools Approach to Public Policy." In *Urban Design in the Real Estate Development Process,* edited by Steve Tiesdel and David Adams, 1–14. Oxford: Wiley-Blackwell.

Tomalty, Ray, and Alan Mallach. 2016. *America's Urban Future: Lessons from North of the Border*. Washington, DC: Island Press.

Tonino, Guilio. 2003. "Consciousness Differentiated and Integrated." In *The Unity of Consciousness,* edited by Axel Cleeremans, 253–65. Oxford: Oxford University Press.

Town and Planning Act 1976 Malaysia. 2005. "Planning Act 1976." In *Compendium of Land Use Laws Development,* edited by John Nolon, 97–111. Cambridge, UK: Cambridge University Press.

Trojan, Przemyslaw. 1984. *Ecosystem Homeostasis*. Berlin and New York: Springer.

Urban Land Institute. 2003. *Mixed-Use Development Handbook*. Washington, DC: Urban Land Institute.

Vernant, Jean-Piere. 1983. *Myth and Thought Among the Greeks*. London: Routledge & Kegan Paul.

Victor T., King. 2008. *The Sociology of Southeast Asia: Transformations in a Developing Region*. Copenhagen: NIAS Press.

Vipond, Robert. 2017. *Making a Global City: How One Toronto School Embraced Diversity*. Toronto, Buffalo, and London: University of Toronto Press.

Voegelin, Eric, and Dante L. Germino. 2000. *Order and History, Volume III: Plato and Aristotle*. Columbia: University of Missouri Press.

Waldheim, Charles. 2012. *The Landscape Urbanism Reader*. New York: Princeton Architectural Press.

Walker, J. H. 2009. "Patrimonialism and Feudalism in the Sejarah Melayu." In *The Politics of the Periphery in Indonesia*, edited by John H. Walker, Glenn Banks, and Minako Sakai, 39–61. Singapore: NUS Press.

Wallerstein, Immanuel. 2011. *The Modern World-System III: The Second Era of Great Expansion of the Capitalist World-Economy, 1730s–1840s*. Berkeley, Los Angeles, and London: University of California Press.

Wheatley, Paul. 1983. *Nagara and Commandery: Origins of Southeast Asian Urban Tradition*. Chicago: University of Chicago, Research Paper of Geography.

William, Raymond. 1958. *Culture and Society*. New York: Columbia University Press.

Williams, Jean Kinney. 2009. *Empire of Ancient Greece*. New York: Facts on File.

Williams, Katie, Michael Jencks, and Elizabeth Burton. 2000. *Achieving Sustainable Urban Form*. London: Taylor & Francis.

Wilson, Edward O., and Frances M. Peter. 1988. *Biodiversity*. Washington, DC: National Academy Press.

Winter, Frederick E. 2006. *Studies in Hellenistic Architecture*. Toronto: University of Toronto Press.

Wirth, Louis. 1938. "Urbanism as Way of Life." *American Journal of Sociology* 44 (1): 1–25.

Wisseman-Christie, Jan. 1991. "States Without Cities: Demographic Trends in Early Java, Indonesia." *Indonesia* 52: 23–40.

World Population Review. 2017. "Toronto Population 2017." Accessed August 19, 2017. http://worldpopulationreview.com/world-cities/toronto-population/.

Xiangming, Chen, and Ahmed Kanna. 2012. *Rethinking Global Urbanism: Comparative Insights from Secondary Cities*. London and New York: Routledge.

Yaakob, Usman, Tarmiji Masron, and Fujimaki Masami. 2000. "Ninety Years of Urbanization in Malaysia: A Geographical Investigation of Its Trends and Characteristics." Working Paper, Ritsumeikan: Ritsumeikan University.

Yap, Kioe Sheng. 2012. "The Challenges of Promoting Productive, Inclusive and Sustainable Urbanization in Southeast Asia." In *Urbanization in Southeast Asia: Issues & Impacts*, edited by Kioe Sheng Yap and Moe Thuzar. Singapore: Institute of Southeast Asian Studies.

Zijderveld, Anton C. 1998. *A Theory of Urbanity: The Economic and Civic Culture of Cities*. Piscataway, NJ: Transaction.

Index

Printed in the United States
By Bookmasters